AF852314

*The Sonic Surrogate*

CALIFORNIA STUDIES IN MUSIC, SOUND, AND MEDIA

*James Buhler and Jean Ma, Series Editors*

1. *Static in the System: Noise and the Soundscape of American Cinema Culture,* by Meredith C. Ward
2. *Hearing Luxe Pop: Glorification, Glamour, and the Middlebrow in American Popular Music,* by John Howland
3. *Thinking with an Accent: Toward a New Object, Method, and Practice,* edited by Pooja Rangan, Akshya Saxena, Ragini Tharoor Srinivasan, and Pavitra Sundar
4. *Key Constellations: Interpreting Tonality in Film,* by Táhirih Motazedian
5. *Just beyond Listening: Essays of Sonic Encounter,* by Michael C. Heller
6. *Making Stereo Fit: The History of a Disquieting Film Technology,* by Eric Dienstfrey
7. *The Composer's Black Box: Making Music in Cybernetic America,* by Theodore Gordon
8. *Automatic Artistry: Negotiating Musical Creativity in a Technological Age,* by Alyssa Michaud
9. *Preprogrammed: How Electronic Presets Changed Music and Media,* by Amy Skjerseth
10. *Sonic Infrastructures: Mediating National Cinemas in West and South Asia,* by Claire Cooley
11. *The Sonic Surrogate: Music, Technology, and the Crisis of Reproduction,* by Marie Thompson

# The Sonic Surrogate

## MUSIC, TECHNOLOGY, AND THE CRISIS OF REPRODUCTION

Marie Thompson

UNIVERSITY OF CALIFORNIA PRESS

University of California Press
Oakland, California

© 2026 by Marie Thompson

All rights reserved.

Library of Congress Cataloging-in-Publication Data

Names: Thompson, Marie, 1987– author
Title: The sonic surrogate : music, technology, and the crisis of reproduction / Marie Thompson.
Other titles: California studies in music, sound, and media 11.
Description: Oakland, California : University of California Press, [2027] | Series: California studies in music, sound, and media ; 11 | Includes bibliographical references and index.
Identifiers: LCCN 2026026750 (print) | LCCN 2026026751 (ebook) | ISBN 9780520433649 cloth | ISBN 9780520433656 paperback | ISBN 9780520433663 epub
Subjects: LCSH: Music—Social aspects | Household electronics—Social aspects | Digital jukebox software—Social aspects | Music and technology | Music therapy | Sleep | Child rearing | Prenatal care
Classification: LCC ML3916 .T53 2027 (print) | LCC ML3916 (ebook) | DDC 306.4/842—dc23/eng/20260713
LC record available at https://lccn.loc.gov/2026026750
LC ebook record available at https://lccn.loc.gov/2026026751

GPSR Authorized Representative: Easy Access System Europe, Mustamäe tee 50, 10621 Tallinn, Estonia, gpsr.requests@easproject.com

35 34 33 32 31 30 29 28 27 26
10 9 8 7 6 5 4 3 2 1

# Contents

# List of Illustrations

# Introduction

## REPRODUCTIVE SOUND TECHNOLOGIES

This book addresses a deceptively simple question: what is sound's relationship with reproduction? On this basis, readers might assume that what follows concerns the capture, repetition, and distribution of recorded sound; the relationship between an original auditory event and its copy; or ideals of sonic fidelity. This, after all, is what "sound reproduction" conventionally signifies. And these concepts, histories, and practices are certainly relevant, but they are only part of the story: sound (in general) and music (specifically) are implicated in other politics, processes, and relations of reproduction.

The motivations for asking this question stem from an apparent renewal of interest in music's relationship to care. A 2018 listicle by Jessica Booth for the online women's magazine *Bustle* is exemplary of this trend. Titled "12 self-care playlist ideas to listen to when you just need time for yourself," the article instructs the reader that "sometimes life gets to be too much to handle." Yet music can enable us to keep going: "its serious impact upon your emotions and body" means that "it's always a good idea to have some self-care playlist ideas on hand that will really make you feel better."[1] The article therefore recommends some Spotify playlists "for whenever you're in need of some extra help." For those who want to relax during a bath, there is "self-care night," featuring tracks by Yo La Tengo, Norah Jones, and Madeleine Peyroux. Spotify's "Happy Hits!" is recommended for when you need a little extra energy, while "Calm Vibes" is suggested for those who want an uninterrupted nap. If your boss has been overly critical, or you are experiencing heartbreak, then the playlist "Fuck them" is a much-needed "confidence booster," featuring songs by pop icons such as Taylor Swift, Ariana Grande, and Rihanna. An untitled trap and hip-hop playlist, meanwhile, is for those who need the relief of a good long cry.

*Bustle*'s listicle is unremarkable. This type of publication is a quotidian feature of the digital media landscape. Nevertheless, its presentation of music (in this instance, mood-oriented Spotify playlists) as a means of enabling and enhancing care (in this instance, an explicitly feminized notion of self-care) is revealing of a recurrent dynamic, whereby music and its technologies of mediation are figured as tools *to help us look after ourselves.*

Music's expediency extends beyond *self*-care. Take Vera: an app launched in 2022 by the wellness and AI startup Music Health. Facilitated by a partnership with Universal Music Group and brought to market with an Australian public-funded research grant, Vera offers personalized music selection that can improve the cognitive functioning and mood of people with dementia by stimulating the part of the brain responsible for long-term memory. It is described by the Music Health co-founder Stephen Hunt as follows:

> We've built Vera to know and find the music that means the most and has the biggest effect for each person living with dementia. It acts like a music detective that seeks out tunes that they used to love a long time ago but may have forgotten about, which their carers may have never heard of, and their families may not even know.[2]

While its algorithmic "music detective" uses personal data to curate a playlist of familiar music for a person living with dementia, Vera is framed as an app *for* carers.[3] And these caregivers, it would seem, need all the help they can get. In the UK, where dementia is one of the lead causes of mortality and later-life disability, formal and informal dementia care systems are recurrently described as at a breaking point. This is largely due to the prohibitive costs in accessing adequate support: a problem that the family, and women family members in particular, often find themselves ameliorating by taking on caring responsibilities themselves.[4]

The Bustle listicle and Music Health's app exemplify how the necessarily broad concept of care has become increasingly enmeshed with recorded music and its digital circulation. Care is also integral to social reproduction. Social reproduction concerns the activities and relations that produce capital's workers and with them, the strange commodity of labor-power. Pertinent for the analyses pursued over the course of this book, it also concerns the maintenance of capital's future and former workers (e.g., children and the elderly) and others "surplus" to capital's demand for labor. Care, which is to say, the nurturing relations that help secure and sustain individual and collective wellbeing, is a key component of social reproduction.

Presently, care is often spoken about in terms of scarcity and strain: the aforementioned "breaking points" are not unique to dementia care systems. Such articulations are expressive of what has been recognized as an enduring *crisis of reproduction*.[5] Associated with the transformations enacted by neoliberalism, post-Fordism, and financialization, this crisis appears as a set of destabilizing pressures impacting the production and care of children, the maintenance of families, households, and other forms of kinship and community, and the availability of "life-making" resources. The crisis of reproduction has at its root the antagonism between labor and capital and the subordination of need to profit-seeking and accumulation. It has a variety of symptoms, including rising rates of ill-health, disablement, and homelessness; loneliness "epidemics"; diminishing access to affordable childcare and specialist social care; declining birth rates; and "time-poor" women.

Music and its technologies of mediation do not simply reflect this crisis: they actively participate in this crisis, promising a cost-effective solution to its effects. In the late twentieth and early twenty-first centuries, what I want to name as *reproductive sound technologies* have proliferated within developed economies, appearing as tools that can enhance and assist in the rearing of (preborn) infants, the management of the household, the facilitation of sleep and the provision of social care, so as to create productive workers, "competitive" children, and "independent" elders. In so doing, these *sonic surrogates* provide a commodified fix to the social problems arising with austerity, the rescindment and recomposition of welfare provision, and the accompanying *externalization* of reproductive responsibility onto individuals and families.

The social provisions reproductive sound technologies promise to intervene in are bound up with gendered stratifications. Women have been and remain responsible for much of capitalism's life-making activities, but race, class, sexuality, and historical circumstance have conditioned how and by whom these activities are performed: gender, after all, is not a static social form. Wealthier women have often outsourced their responsibilities onto poorer women acting as nannies, domestic servants, and cleaners, and this outsourcing has only intensified as women have (re)entered the workforce over the late twentieth and twenty-first centuries. The externalization of reproduction, enacted by making individuals and families (rather than states and corporations) increasingly responsible for fulfilling their own needs, has been accompanied by the expanded *commodification* of caring services to be paid for by the wage. Often performed by low-paid migrant workers, these goods and services—from private nurseries to the food delivery and cleaning services of the "gig" economy—enable those that can afford it to ameliorate some of the demands of care and the household.[6]

Expressive of reproduction's externalization and commodification, reproductive sound technologies are gendered—and gendering—technologies. To this end, their design and circulation often references—but sometimes scrambles—gendered aesthetics, practices, and formations. At the same time, reproductive sound technologies reflect a desire for the automation of gender, for the machine to undertake some of the caring, maintenance, and "gestational" labors associated with women: as mothers and housewives but also as members of a feminized and racialized workforce onto whom reproductive responsibilities are outsourced.

There is, however, another side to reproductive sound technologies. Social reproduction concerns not only the reproduction of the working class and those externalized by the wage. It also concerns the reproduction of capitalist society's constitutive relations. As a result, capitalist social reproduction is not simply adopted but enforced. Techniques of enforcement have taken the form of violence, juridical legislation, religious treatise, and carceral expansion.[7] Consequently, I understand reproductive sound technologies to refer not only to devices that promise to help us fulfil our *reproductive responsibilities* but also those that police, enforce, and maintain *reproductive respectability*. Reproductive sound technologies can be disciplinary or punitive: they affirm, remediate, and in some instances, intensify capitalist norms of a "respectable" life and prohibit deviation from these norms. Indeed, the latter function is implicated in the former: the adoption of capitalist reproductive respectability, upon which reproductive sound technologies are modeled, forecloses other possible modes and means of life making.

Imbricated with shifting formations of capitalist production and social reproduction, reproductive sound technologies reflect correlate changes within auditory culture. This includes the temporal and spatial expansion of music consumption facilitated by the proliferation of speaker systems (from PA announcement systems to "smart" headphones); the convergence of the music industry with the digital economy; the rise of music streaming and the shift from commodity ownership to rented access on the part of the consumer; and the "personalization," "curation," and "datafication" of music listening.[8] However, the relationship between sound reproduction, care, and social reproduction long precedes these transformations. Since the late nineteenth century, sound reproduction technologies have been shaped by gendered, racial, and colonial divisions of labor upon which capitalist production and reproduction rely. Furthermore, as I foreground in chapter 1, the early commodification of sound's mechanical reproducibility established the notion that recorded music and its associated technologies may act as sonic surrogates, reducing or even replacing "mother's" work. While the technological fixes offered by music goods and services in the late twentieth and twenty-first centuries are conditioned by the externalization and commodification of reproduction within developed economies, as well as the logics of neoliberalism and financialization, they are also conditioned by a longer history of sound's technological reproducibility.

## REPRODUCTION'S RESONANCES

*The Sonic Surrogate* engages with the histories and discourses of sound reproduction via the feminist problematic of social reproduction. Yet enmeshed with these two, seemingly disparate conceptual and theoretical lineages are other notions of reproduction. As Alys Eve Weinbaum outlines, in its relatively short history, *reproduction* has accrued a variety of meanings. Originating as a synonym for "resurrection" in the work of seventeenth-century theologians, it was only in the latter half of the eighteenth-century that reproduction came to mean species reproduction.[9] As a biological and sexual process, reproduction is considered fundamental to the continuation of life. Reproductive institutions—of race, nation, and family—are organized around fears and aspirations of how this "life" is continued and are bound up with notions of inheritance, legacy, legitimacy and eugenics.

If social reproduction concerns the historically, culturally, and socially specific activities, relations, resources, and formations that produce and sustain capital's laboring classes and those not directly involved in the formal economy, then this involves the biological—or more accurately, biosocial—processes of pregnancy, gestation, and childbirth. Shifting configurations of race, nation, and family, furthermore, have been central to the organization of life-making activities within capitalist society. Yet despite their social necessity, these activities have often appeared contingent, insignificant, or external to economic concerns: outcomes of love, natural proclivity, or personal desire. Consequently, Marxist feminist activists and theorists have drawn attention to the what Leopoldina Fortunati names the "arcane" of reproductive labor: the oft-concealed, "naturalized," and largely nonmarket-mediated domain that, in its production of labor-power, is key to capital accumulation.[10] Reproduction was politicized as a terrain of struggle by movements such as Wages for Housework, which emerged in the 1970s: if the wage was a designation of productive labor, then the demand of wages *for* housework acknowledged its centrality to "all industry and profit."[11] Yet, for many involved, this demand was not to be interpreted literally. Rather, it sought the expansion of the category of productive labor for the purpose of struggle *against* capitalist exploitation, as Nicole Cox and Silvia Federici clarify: "By saying that the work we perform in the home is capitalist production, we are not expressing a wish to be legitimated as part of the 'productive forces,' . . . Ultimately when we say that we produce capital we are saying that we can and want to destroy it."[12]

In considering how social reproduction relates to accumulation, feminist analyses have provided an important corrective to its historical exclusion from understandings of capitalist production. These analyses have continued with the recent expansion of interest in "social reproduction theory," building upon but also diverging from earlier feminist accounts and addressing different histories, social practices, institutions, and political struggles relating to life making in and at the margins of capitalist societies.[13] However, as a result, the "reproduction" of social reproduction has itself come to hold different meanings, sometimes resulting in ambiguity. There are important tensions between the treatment of reproductive labor as a discrete, gendered set of tasks, activities, and roles and social reproduction as a structural formation consisting of social relations, processes, and practices deemed necessary for the reproduction of laboring populations, including public education, health care systems, community leisure facilities, housing, pensions, and benefits. Whereas the latter approach risks indeterminacy, inasmuch as it is difficult to distinguish where the limits of reproduction lie, the former risks being too restrictive, inasmuch as reproductive labor becomes conflated with general—and sometimes valorized—notions of care or housework.[14] There are also ambiguities that arise when reproduction can simultaneously refer to unremunerated, informally paid, and formally waged activities (there are, for example, important differences, as well as connections, in the material conditions of low-paid caregivers and unpaid family carers), as well as referring to the reproduction of labor and the reproduction of capitalist social relations.[15] This book works across these ambiguities, addressing reproduction as it pertains to unwaged, "natural" activity, an exploited feminized workforce, and the production of social stratifications.

Of particular significance here is an identifiable shift in the organization of capitalist social reproduction within developed economies, the trajectory of which can be generalized as follows.[16] The distinction between the spheres of production and reproduction, to which 1970s and '80s Marxist feminism tends to refer, characterizes the compact of the Fordist family wage and, before that, the "housewifization" of women in the industrial, colonial core (see chapter 1): work and home were recognized as separate domains, corresponding to the breadwinning man and "motherly" wife, respectively. The separation of the two spheres was never complete: poor and working-class women often worked outside the home (or performed paid "outwork" inside the home), while the Fordist compact was restricted to white, unionized men.[17] This compact belonged to a regime of state-managed capitalism that emerged out of the Great Depression and Second World War, characterized by a combination of industrial production and domestic consumerism (both of which were enabled by ongoing colonial and postcolonial expropriation) and the expansion of state and corporate provision for social welfare, with public investment in healthcare, education, and housing.[18]

Over the past four decades, the constitutive relations of state-managed capitalism and the Fordist family wage have been slowly and substantively recomposed. "Stagflation" and the "productivity crisis" of the 1970s bolstered neoliberal initiatives to expand marketization. Post-Fordism established workforce "flexibility," the fragmentation of production, deindustrialisation and the relocation of manufacturing to low-wage regions, while financialization saw "fictious capital," assets, and speculation take on increasingly central roles within society and economy. Amid these changes, state and corporate welfare is rescinded and reorganized in accordance with market logics, and the ideal of the two-earner family emerges. This is accompanied by the aforementioned externalization and commodification of reproduction: the return of reproductive responsibility onto the individual and family, so that needs are met by the private household rather than through social provision or employer; and the correlate expansion of marketized goods and services.[19] At the same time, the rewards of waged work are increasingly insecure: employment becomes temporary and fragmentary, while wages stagnate and fall, necessitating the expansion of the working day to try to maintain a liveable life. Indeed, a growing number of the global population finds itself external—or "surplus"—to the wage-relation and "exists now only to be managed."[20] As it becomes increasingly difficult to keep up with the costs of reproduction, debt takes on new disciplinary and accumulative power: providers of credit are increasingly used to supplement work, welfare, or both, so as to enable the continuation of consumer spending.[21] The destabilizing effects of these social and economic transformations, I contend, provide the background conditions for many of the reproductive sound technologies discussed here.

While there is a rich body of scholarship concerning music's relationship with capitalism, attention to social reproduction within it has been scant.[22] This omission is notable since reproduction has long had an auditory component: practices such as the lullaby (see chapters 1 and 5) are exemplary of how the gendered division of labor has also been a gendered division of *musical* labor. While what is pursued in this book is not a corrective analysis of the social reproduction of music's formal and informal workforces, it nevertheless makes apparent some of the ways in which reproduction has been missed. For instance, the temporal, spatial, and affective binaries through which histories of sound recording and reproduction are commonly narrated—labor/leisure, work/home, drudgery/entertainment—relate primarily to the world as it appears for the archetypal (white, masculine) industrial waged worker and therefore struggle to account for the frequently feminized household responsibilities that are not quite work and not quite leisure either. Indeed, reproductive sound technologies destabilize the work-leisure distinction that has provided the ground for dominant narratives about sound technologies and their uses: by using sound and music to assist in the social labors of life making, reproductive sound technologies promise something more, or something other, than entertainment.

I thus approach reproduction, its technical, social, and biological connotations, and the resonances between them as a heuristic for understanding, connecting, and explicating the ways in which auditory technologies are posited as surrogate mothers, servants, carers, and housewives that help us around the home (chapters 1 and 2); as providing types of rejuvenation and relief that enables us to be more efficient, effective, and successful in life and work (chapter 5); as a force in the revanchist policing of loiterers, the homeless, and other urban surplus populations (chapter 4); or as improving current and future children (chapter 3). I begin from the assumption that, as Jonathan Sterne argues, "sound-reproduction technologies are artefacts of particular practices and relations 'all the way down.'" If, as Sterne observes, the "torrential rains of capitalist development and marketing that pelt us with new digital machinery" make it "both easy and tempting to forget the enduring connection between any technology and a larger cultural context," a critical reappraisal of the reproduction of "sound reproduction" helps to foreground the relations between media technics, hierarchies of gender, race and disability, and social divisions of labor.[23]

## CRISIS, AUTOMATION, AND THE FAMILY FORM

Three connective threads run through this book: *crisis, automation,* and *the family form.* These concepts recur across the chapters, mediating the relationship between sound and reproduction in different social and historical environs.

Bookended by two crises within the music industry—the crisis of musical labor stemming from the growing availability of sound reproduction technologies in domestic and public culture in the early twentieth century and the economic crisis induced by the emergence of peer-to-peer networks nearly a century later—the reproductive sound technologies considered in each of the chapters pay testament to various social, economic, and aesthetic instabilities. These include enmeshed crises of the 2008 financial crash, austerity and social care (chapter 2), crises of childhood and childrearing (chapter 3), crises of housing, homelessness, and antisocial behavior (chapter 4), and the crisis of sleep (chapter 5). Many of these crises can be traced to the destabilizations of capitalist production and reproduction that characterize the post-Fordist period. They are also components of what I am strategically referring to as a crisis of reproduction. Such a characterization, however, requires some qualification.

If the historical designation of this crisis appears somewhat tenuous, then this partly due to a crisis tendency common to all capitalist modes of social reproduction. As Nancy Fraser makes clear, far from being exceptional events, these various, interlinked, and protracted crises of "care" are symptomatic of the capitalism's tendency to become unmoored from its own social base.[24] Put bluntly, they are a feature, not a bug. For some feminist commentators, including Fraser, this crisis tendency stems from a fundamental contradiction between social reproduction and capital. This is because social reproduction is fundamental to capitalist relations, providing value-generating labor power, current and future workers, and the reserve army of labor. Yet social reproduction also acts as a limit, placing a break on unfettered accumulation. It is, in other words, a necessary constraint.

This line of argument, however, risks assuming the destabilization of capital accumulation will necessarily follow from the destabilization of social reproduction. As Smriti Rao suggests, there is an important distinction to be between a crisis of reproduction for *labor* and a crisis of reproduction for *capital*. The latter involves changes to the quantity or quality of labor power that threaten to destabilize accumulation, while the former manifests as the household, family, or community struggling to secure what is needed for a liveable life.[25] The crisis in which much of the world's working class struggle to fulfill their needs, is not "a crisis *of* capitalism" per se: rather, it demonstrates how the fulfillment of these needs is made dependent upon the ups and downs of capitalist economy.[26] Indeed, as will become apparent in chapter 2, which details the incorporation of Amazon's smart speaker into the UK's local authority social care provision, what, from the perspective of labor, appear as various, interlinked crises of care have, from the perspective of capital, provided opportunities for profit seeking.

Automation has often appeared as both the cause of and solution to capitalism's crises. Technological interventions that reduce or replace human labor may condemn many to unemployment; alternatively, they promise a world where the demands of work are minimized, enabling more free time. These are characteristic propositions of what Aaron Benanav diagnoses as "automation discourse": a recurring discursive tendency within capitalist societies that typically emerges in times of labor market instability. It has reappeared, once again, as the low demand for labor (exemplified by stagnating wages, rising inequality, and spikes of unemployment) is accompanied by a new wave of automation technologies, primarily driven by artificial intelligence.[27] For Benanav, automation discourse mistakenly establishes causality between the latter and the former: it presents technological change as the driver of shrinking labor markets, therefore downplaying the role of persistent economic stagnation and low rates of global growth.[28]

Automation discourse has typically been tethered to questions of productivity and access to the wage. As a result, comparatively less attention has been afforded to automation on the terrain of reproduction. Yet from the washing machine to robotic vacuum cleaners, from the phonograph to the smart speaker, consumer technologies have been presented as "labor-saving" interventions that can reduce the time and effort expended on reproductive labor. Furthermore, the expansion of the digital tech sector, and its convergence with the service industries, has bought a wave of technological interventions and mediations facilitating the apparent automation of gender. Enabling the outsourcing of reproductive labor onto an app and onto the precarious service workers it summons, these interventions belong to what Sarah Sharma dubs "the post-mom economy," capturing a particular moment in male-dominated tech culture, of which Uber ("Mommy, drive me"), TaskRabbit ("Mommy, clean my room"), GrubHub ("Mommy, I'm hungry"), and LiveBetter ("Mommy, I'm bored") are emblematic. These products promise to "dispense Mommy" without "dispensing *of* Mommy": they expand, intensify, and commodify Mommy's reproductive responsibilities, attaching them instead to variously gendered gig workers and digital platforms, while also maintaining a reliance on Mommy to fulfill our needs: "if Mommy cannot or will not provide that care anymore, perhaps a new machine can—or, if not the machine itself, a marginalized other summoned via machine."[29] While the reproductive sound technologies considered in this book are less reliant on precariously employed gig service workers than the platforms Sharma names, they reflect a similar desire for "mothering machines": technologies that care for us, enabling a cheap and simple fix to the demands associated with self-maintenance, family, and the home.

Strictly speaking, automation describes machines that fully replace human labor. Distinctions are frequently drawn by theorists of work technologies between mechanizing, labor-augmenting devices that increase worker productivity (therefore resulting in the need for fewer workers) and automating, labor-substituting technologies (which dispense of the need for the worker entirely). Yet, as Benanav notes, in practice, distinguishing between these two types of "labor-saving" interventions and between change in a labor process and the elimination of a job role can be challenging.[30] This is, in part, because automation technologies do not simply replace human labor: they interact with labor processes in complex ways. Just as the service apps referred to by Sharma that appear replace "mom's" work with the click of a button are in fact reliant on the exploitation of low paid, precarious gig workers, what appears as the work of the machine remains reliant on human labor chains. As Gavin Mueller argues, automation "recomposes the workforce, isolating and rearranging tasks, and hollowing out middle-tier occupations." Certain types of work process remain too costly or too challenging to automate, particularly when the costs are evaluated in comparison to cheap human labor. The reliance upon a "stratum of human labor that is all but ineradicable" means that automation invariably stops short of being "full."[31] Reproductive sound technologies, for their part, are constituted by transnational networks of exploitation, extraction, and accumulation, connecting the human labors implicated in the mining of natural resources, electronics, and hardware manufacturing, and the composition, production, and reproduction of music. When they manifest as digital or "smart" technologies, they are driven by the hidden "ghost work" of data processing that enables the internet and its subsidiaries to appear intuitive, responsive, and efficient.[32]

The difficulty in distinguishing mechanization from automation, augmentation from elimination, and labor's reduction from its recomposition is exacerbated when technological surrogacy concerns reproductive labor. As discussed in chapter 1, there are limits to describing reproductive sound technologies as labor-saving devices, insofar as reproductive labor has a distinct relationship to time and measurability that means little of this labor can be "saved." Indeed, social care's smart speakers (chapter 2) or mental well-being apps (chapter 5) both speak to care's resistance to technological "improvements" in productivity, insofar as trying to make labor-intensive care work more "efficient" tends to result in a lower standard of care.[33] Here, too, we might recall the Vera app, promising to relieve some of the burden on caregivers. At play is a distinction between producing goods such as smartphones and speakers, which can be done relatively cheaply, and the "high-cost, low-wage" care sector: reproductive sound technologies offer the prospect of reducing the need for the latter (high-cost care) with the former (low-cost music and its technologies).[34] Indeed, as Benanav and Clegg have argued, the transformation of services (which, like care, are labor intensive) into consumer goods has been a key means through which the former have been made more profitable for capital since, unlike services, the production process of goods is amenable to what Marx calls real subsumption (i.e., the labor process can be organized to meet capital's own ends by enabling increases in productivity, typically through the adoption of technological efficiencies in the production process).[35] While some reproductive sound technologies (e.g., prenatal sound systems) commodify the "free services" of maternal responsibility via consumer goods, others (e.g., smart speakers and musical loitering deterrents) supplement or stand in for the low-paid services of carers and private security guards.

Increased efficiency, however, is only part of what drives automation: as Mueller suggests, greater control (for the capitalist) and reduced autonomy (for the laborer) over the labor process is also at stake.[36] It is this tendency that helps explain the capitalist rationale for reproductive sound technologies. Not only do they represent an expanded field of consumption and the perpetual generation of novel lifestyle products. As music's mediation becomes increasingly inextricable from digital infrastructures, enabling it to become a technology of surveillance, reproductive sound technologies also contribute to the blurring of the distinction between production and reproduction: playing music to fetuses, reporting antisocial noise in the local area, or selecting a soundtrack to fall asleep to become opportunities for informational extraction and accumulation.[37] Thus, the automation performed by reproductive sound technologies, particularly when facilitated by the digital economy, marks capital's drive to enter, organize, and extract profit from previously noncommodified times, spaces, and relations.

The heterosexual, cisgendered, and white-racialized construct of the private nuclear family has endured as a core component of capitalist social reproduction. While the norms of this compact have undergone a series of historical shifts into the twenty-first century, the family has remained a primary provider of care and life-sustaining resources within capitalist society. Thus, it is the family that has provided—and is assumed to provide—an (increasingly threadbare) safety net to the crisis of reproduction: familial obligation drives its members to balance the demands for care with the need to acquire economic resources. Of course, not everyone experiences the family as a caring institution. As M. E. O'Brien observes: "The family is a joy to some, a necessity for most, and a nightmare for too many. Behind its closed doors, the household is a gamble."[38] The family can be a site of violence, as well as a target of state violence, particularly as the boundaries between health care, social services, and the criminal justice system become increasingly porous (see chapter 4). It is both an oft-romanticized ideal that is difficult to shake off and a coercive and disciplinary compact that mediates the demands of life and work.

The family form is also integral to histories of sound reproduction and continues to be a key mediator of how audio technologies are designed, used, and socially situated. Yet the persistence of the family within auditory culture, particularly within the late-twentieth and twenty-first centuries, has frequently been missed. Instead, accounts of "neoliberal" sonic media have typically centred on the individual. The promises of personalization and self-control offered by music streaming services and their associated technologies are taken as emblematic of neoliberalism's treatment of the atomized individual as the primary economic and political unit.[39] Yet what this omits—and what, I contend, reproductive sound technologies reveal—is that neoliberalism continues to value and valorize the family. Here, I follow the lead of Melinda Cooper, whose incisive historical examination of the convergence of US neoconservatism and neoliberalism makes apparent how the family remains a foundational compact. Although neoliberals may exhort individuals to "take responsibility for their own fate," as Cooper argues, "the imperative of *personal responsibility* slides ineluctably into that of *family responsibility* when it comes to managing the inevitable problems of economic dependence (the care of children, the disabled, the elderly or the unwaged)."[40] Reproductive sound technologies exemplify the elision of individual and familial responsibility: devices that appear to individuate musical experience by adapting to a listener's tastes, requirements, and requests, enabling them to "soundtrack their lives," to reference Spotify's advertising slogan, also appear as devices to entertain, soothe, and educate children, to manage the household, and to take care of elder relatives.

The omission of the family from theorizations of neoliberal sound technologies is striking given that it has been hiding in plain sight: the proliferation of family plans offered by music streaming services since 2011, for example, is a testament to its proximity to ostensibly "personalized" sonic media.[41] Typically offering six accounts for a reduced subscription fee, the model of the family plan represents the formalization and monetization of informal practices of account sharing, in which a single account might be shared between multiple friends and family. They enable the pooling of resources between a household by enabling multiple members of a family to simultaneously access the private library of a streaming service, while also maintaining the algorithmic "purity" of each family member's account so that their search history and recommendations are unaffected by other household listeners. In many instances, family plans are associated with features that extend the surveillant capacities of the platform to parents or the legal guardians of children, so that they are able to monitor and regulate their listening. Specialist features such as Spotify Kids, an app launched in late 2019 for those aged three plus, offers "kid friendly" media to "start them off on the right track" and take them "from playtime to bedtime," while also allowing parents to set "age-appropriate" content restrictions and to view what children are listening to. If neoliberalism recognizes children as, in the words of Chicago School economist Milton Friedman, "responsible individuals in embryo,"[42] where responsibility is primarily defined in terms of consumption (i.e., the use of economic resources as one sees fit), then streaming services enable children to practice the exertion of choice within the constraints produced by parental regulations. However, these parental regulations are a matter for the private family, insofar as "parenting is personal," to quote Spotify.[43] The "family" of the family plan, then, is a distinctly neoliberal construct: a composite of free individuals and free individuals in development, a private institution that is socially and economically responsible for its members and through which "the next generation of listeners" is produced.[44]

If reproductive sound technologies offer to help the family as it is increasingly pressured by the crisis of reproduction, they also step in when the family "fails." As is made apparent in chapter 4, those families that struggle to adhere to ideals of reproductive respectability and reproductive responsibility have faced disproportionate scrutiny from political and legal systems. As bad parenting, and specifically bad mothers, have borne the blame for antisocial behavior and other emblems of "decline," music and its technologies have become a means of offsetting these failings, defending neighborhoods and urban spaces on behalf of "hard-working families." Reproductive sound technologies, then, exist as a means of administering familial relations of care, as well as a means of policing and defending normative kinship relations.

Taken together, crisis, automation, and the family form are integral to the proliferation of reproductive sound technologies. They draw together the seemingly disparate settings, practices, and histories covered here, aiding the transitions between the Taylorist and neoliberal home, revanchist and postrevanchist urban spaces, the uterus and the bedroom. As these different auditory sites intimate (and as is likely already clear from this introduction), I take a necessarily capacious approach to citation: while sound and music studies and feminist accounts of reproduction provide the foundations of this book, it is also indebted to disciplinary contributions from media studies, human geography, anthropology, and cultural history. Satirical cartoons, corporate patents, industry press releases, adverts, and promotional videos all find their place here, as do policing handbooks, housing policies, and local authority ombudsman rulings. Acting as a series of refrains, crisis, automation, and the family form ground what follows.

## MAKING MUSIC USEFUL

The following chapters examine many elements of auditory culture: from the noise of surplus populations to the listening work of domestic servants. In this regard, *The Sonic Surrogate* is reflective of tendencies within a broader auditory technoculture, in which music is one form of sonic media among many, appearing alongside soundscapes, audiobooks, and guided mediations. Indeed, if the separations of production/reproduction, labor/leisure, home/work appear increasingly precarious, so too are those that distinguish sound from music. At the same time, it is undeniably music, and specifically recorded music, to which the sound of reproductive sound technologies primarily refers. These devices produce a particular figuration of music as useful. As will become clear, their articulations of music's usefulness are often mediated by the imperatives of the private family, state, and capital. Music is useful in that it diverts and provides companionship, alleviating loneliness and entertaining those in need of care; and it is also useful due to its ability to impress upon listeners, modulating their psychic, cognitive, and emotional state. These technologies are therefore indebted to—and reinforce—a wider instrumentalization of music as a therapeutic tool capable of fulfilling particular social, subjective, and affective needs. This renewed emphasis on music's expediency is reflected in the organization of its contemporary media infrastructure: as the self-care playlists discussed at the beginning of this introduction exemplify, many streaming services make music available for consumption in accord with mood or activities, while smart speakers and smart headphones are designed for integration within daily routines, enabling uninterrupted connection to music's digital platforms.[45] It is also revealed by the partnerships between music recording companies and health, fitness, and well-being businesses; the convergence of the music tech and health tech sectors; and the widespread adoption of music by the third sector as a means of engaging with "at-risk" populations (see chapter 5). Such presentations of music are undoubtedly connected to compelling yet often dubious ideological investments concerning the medium's particular affordances—its ability to reach where language cannot, to connect across social barriers, and to alter body and mind—but also the very real intimacies, socialities, and pleasures implicated in music. If music is often experienced, shared, valued, and enjoyed as a "world-shaping" medium, then reproductive sound technologies call upon and reaffirm its status as such by putting music's affects and significations to work.

It is music's cheapness, as well as its usefulness, that lies behind its presentation as a fix for the crisis of reproduction. What Timothy Taylor succinctly refers to as the "ideology of plenitude" has long accompanied music's commodification and reification.[46] However, as Eric Drott makes clear, the rise of streaming services has served to intensify music's status as a cheap and abundant resource: metaphoric comparisons to water, for example, speak to a situation where accessing a seemingly boundless supply of music is as simple as turning on a tap. While its cheapness makes it an appealing intervention for the purposes of care and other types of social reproduction (a point that will recur across this book), this hinges on the suppression of the cost of music, including the costs of musical labor. At play here is music's partial resistance to real subsumption, which has contributed to a romanticization of musical labor as that which is undertaken for reasons other than remuneration, and the "oversupply" of musical labor due to relatively low barriers to entry and the perceived desirability of this work. Yet music's mechanical and digital reproducibility plays a role too, reducing both the need for labor and costs of music production. Indeed, recorded music can itself be understood as part of capitalism's transformation of lower-profit services into higher-profit consumer goods, one which enables greater productivity and lower labor costs: "True, recorded music isn't a perfect substitute for live musical performance. But in many cases, it's a perfectly adequate substitute."[47] For Drott, therefore, the use of music to fill the gaps left by the crisis of care it is shadowed by another crisis within the world of music, as fewer musicians can sustain themselves through their activities.

To acknowledge music's cheapness should not be taken to ascribe it a lack of profitability: music's usefulness has various uses for the different actors that constitute the music industry. For streaming services, music's mediation in accordance with different moods, functions, and contexts are one of various means for keeping listeners listening within the proprietary space of the platform and contributing to the production of information about service users. For recording companies, the attachment of music to function can generate licensing agreements and help accumulate plays of their music catalogue; it can provide a way of connecting recordings to listeners, helping artists to gain audibility amidst the vast music libraries of streaming services.

There is, of course, nothing new per se about the instrumentalization of music: across many histories and cultures, it has been envisioned a functional medium for different ends, and as chapter 1 makes clear, the notion that recorded music can be used to produce emotional, psychic, and social outcomes developed alongside the phonograph. To this end, commentators have countered crude presentations of music's digital culture as divorced from analogue predecessors by drawing attention to the significant resonances with the twentieth-century history of Muzak.[48] A specific brand of functional music, Muzak sought to employ the formal attributes of music to enhance productivity in the workplace and, later, consumption within consumer space. In treating recorded music as a means of attuning the listener, modulating their affective state and regulating their bodily rhythms, it prefigured many of the ways that music is treated as an ambient sonic accompaniment today. According to this view, the musical regulation of the Fordist factory line morphs into the white-collar worker's "lo-fi beats" on noise cancelling headphones.[49] Indeed, while specifically composed mood music certainly persists in the twenty-first century, the boundary between Muzak and recorded music more generally has become increasingly blurred.[50] While the golden-age of Muzak is associated with aesthetically bland, rationalized arrangements aimed at a general listener, various types of recorded music, including music historically associated with high culture, can now take on multiple (and sometimes divergent) affective, psychic, and social functions. Hence recordings of Mozart's music can be used to facilitate a study session, enhance a fetus's cognitive functioning or deter antisocial behavior (see chapters 3 and 4). Muzak also differs from the current regime of musical expediency insofar as it was primarily intended for consumption in shared workplaces and commercial settings. It is therefore unsurprising that few (if any) studies of Muzak and its legacies engage with the family and the house as a site of labor: they are, at best, resigned to passing acknowledgements of the adjacent forms of mood music targeted at women in the home.[51]

It is the self-fashioning listening subject that is typically envisaged as the principal successor of Muzak's industrial listener. Emerging from the post-Fordist blurring of home and work, labor and leisure, this subject uses music and its affective resonances as a means of adapting, enhancing, and optimizing the self.[52] Music's *listener* and music's *user* are therefore imagined as one and the same. Conversely, music's expediency for the purposes of reproduction, with which it appears as a tool for taking care of ourselves *and* others that are dependent upon us, destabilizes the oft-conflated figures of user and listener. To this end, this book raises questions as to how the categories of user and listener function when recorded music's expediency becomes a family affair: a means of not so much managing the self but the household. Who is the user and who is the listener when a mother uses recorded music to soothe an infant to sleep or to lighten the mood between warring siblings? Who is the user when music is intended to make someone easier to care for? Likewise, how do we understand the relationship between music's users and listeners when it is used by store owners and travel companies to deter loiterers? Are these figures one or the other, both or neither?

In such contexts, music is not only useful insofar as it works *upon* listeners. Rather, music also performs work *for* users: its usefulness is directly or indirectly modeled upon and understood in relation to the tasks and activities that are, or were, otherwise performed by people. Attending to music's automation (which is to say, the automation of music, and music as automation) makes apparent the connections between music's affective dimensions, on the one hand, and the gendered division of labor on the other, insofar as the management and modulation of mood, by musical means and otherwise, has often been women's work. These convergences are implicit in some commentators' astute descriptions of Muzak as a domesticating medium: Ronald Radano, for example, suggests that music became familiar as an auditory background through home sound technologies such as the phonograph and radio, and so Muzak represents a public extension of private, personal, and domestic experience.[53] Simon C. Jones and Thomas G. Schumacher, likewise, propose that "the stylistic regularity and harmonic simplicity of Muzak suggest a secure, private, domestic world that signifies the comfort and security of the home in terms of a particular bourgeois conception of domestic well-being. . . . It's aim is precisely to make one 'feel at home' whether in an office, factory of airplane."[54] For Jones and Schumacher, Muzak is ultimately a feminized and feminizing medium, serving to soften the edges of boredom and discourage other forms of collective activity in the workplace. Neither of these accounts, however, offers much further interrogation of functional music's domestic resonances. Indeed, the domestic is figured as the other side of work, a symbolic construct or a site of leisure, omitting the social relations through which domestic comfort is constituted. Consequently, while the story of functional music typically begins with artistic predecessors such as Erik Satie's *Furniture Music*, or with Muzak's inventor, General George Owen Squier, here I begin with the domestic musical labor of daughters, mothers, and nursemaids.

Referring to music and its associated technologies in terms of functionality and expediency risks affirming a crude model whereby consumers, audiences, and "users" are simply assumed to adopt the practices, norms, and values adopted and promoted by industry actors. It is certainly true that "actual" users depart from or are suspicious of the proposed uses of reproductive sound technologies, and their cultural reception has sometimes been hostile. Recall, for example, music streaming services' family plans, which allow a household to collectively access a platform's library and features. Some streaming services have sought to address the family as a single, composite listener, by providing playlist functions that synthesize the most listened-to tracks of each member: these include Pandora's Our Soundtrack feature and Spotify's Family Mix. However, some of the media reception of these features, far from conjuring the image of a family happily listening to music together, highlights the potential embarrassment of sharing private listening amongst family members. A *Guardian* article, for instance, conjures the "chaos" of "mum's Zumba hits" colliding with "your Megan Thee Stallion obsession."[55] Furthermore, family plan providers often reference normative family structures in their terms of service by, for example, stipulating that the primary account holder should be the parent or legal guardian of any children included in the plan, and that all plan members should live in the same household. Yet a common line of questioning on online forums is the extent to which it is possible to share a subscription with non-cohabiting family members and friends.[56] For many, it would seem, the family is not the ideal construct through which to pool resources, musical or otherwise.

Although it sometimes addresses questions of reception, ethnographic literature, and user testimonies, the focus of this book undoubtedly centers on the design, circulation, and cultural imaginaries of reproductive sound technologies and offers little comment on "actual" users. Instead, it engages with these devices as "promissory" technologies, oriented toward a future that will be better, or at least easier. While for some users these auditory interventions might well soften the edges of the crisis of reproduction, their grounding in the privatization of need and in a social compact that necessitates competition, scarcity, and exploitation means their promises ultimately remain unfulfilled. Indeed, the *poverty* of reproductive sound technologies and their cheap solutionism necessitates the politicization of both music and reproduction as terrains of struggle. To that end, this book seeks to highlight the frictions, tensions, and contradictions that accompany these devices, remaining open to the different, and sometimes conflicting, ways that they might be adopted. I address multiple counterexamples (from queer-feminist uterine concert halls to slow jam rest mediations) that engage with reproductive sound technologies in ways that run against the norms and conventions of their design and circulation. While I stop short as describing these practices as resistive, they nevertheless offer some refusal of the logic of auditory usefulness as determined by the needs of capital and state, making apparent the (potential and actual) implication of sound, music, and their associated technologies in other modes of sociality and intimacy. It is, after all, too easy to mistake capitalism's drive toward totalization for totality: while often constrained by reproductive respectability and reproductive responsibility, and undertaken out of necessity rather than desire, other (often nascent) forms care, life making, and maintenance coexist.

## CHAPTER OUTLINE

The five chapters of *The Sonic Surrogate* each deal with a specific type or class of reproductive sound technology. Chapter 1, "Mothering Machines: Phonography, Housewifization, Mechanization," begins with a technology typically afforded a starring role in sound reproduction's origin stories: the phonograph. Narrating the housewifization of recorded music's mechanical reproduction in the early twentieth century, I highlight the figuration of the phonograph as a domestic aid and maternal surrogate for the bourgeois home, with which the mechanization of music becomes a means of mechanization. Yet as the phonograph became an increasingly affordable consumer good, it took on new lives outside the bourgeois home. Addressing the imbrication of housewifization with colonization, I examine how the phonograph exploited—and troubled—colonial regimes of extraction and exploitation.

Maintaining chapter 1's focus on the home, chapter 2, "'Hear for You': Smart Speakers, Extractive Listening, and Techno-Austerity," examines the smart speaker and its digital assistant. Focusing on Amazon's Echo, I analyse the smart speaker as a "listening" device modeled on the auditory relations of feminized work. Yet smart speakers are not reducible to their gendered fantasies, domestic imaginaries, and promises of automation. An attention to listening exposes the limits of conceptualizing these technologies in relation to feminized reproductive labor, while also making apparent the alignment of the smart speaker's gendered presentation with underlying economic logics. Noting cultural anxieties about technologies that listen, I identify three intersecting factors that have driven adoption of the smart speaker: first, the cheapness of these technologies; second, the presentation of surveillance as service, aided by their feminization and the naturalization of aurality; and third, austerity and its resulting crisis of social care. Referring to trials of Amazon's Echo devices by English Local Authority care providers, I suggest these devices are markers of "techno-austerity," promising an inadequate yet lucrative fix to the obligation to do more with less.

Where chapters 1 and 2 predominantly concern the private household, in chapter 3, "'Your Womb, the Perfect Classroom': Prenatal Sound Systems and Uterine Audiophilia," turns to the "first home" of the uterus, interrogating the reproductive technopolitics of prenatal sound systems. Situating these devices in relation to neoliberal theorizations of human capital, I argue that in seeking to provide environmental enrichment for a fetal "listener," prenatal sound systems make audible a particular figuration of pregnancy that combines divergent notions of maternal responsibility and passivity. In seeking to "improve" the preborn future child through sound and music, prenatal sound systems enact a biopolitics of impressibility. However, where the fetal body is presented in terms of plasticity, transformability, and the capacity to listen, prenatal sound systems figure the pregnant body as a resonant container. Consequently, these devices exemplify what I outline as *uterine audiophilia*. Highlighting the coherence and tensions between sound studies' "gestational" discourse and the fidelity discourse accompanying prenatal sound systems, I contend that uterine audiophilia serves to disappear the broader matrix of reproductive relations.

Chapter 4, "Auditory Revanchism, Antisociality, and Reproductive Respectability," addresses the enmeshment of care and criminalization by turning attention to "revanchist" reproductive sound technologies. Building upon geographic discourses of urban renewal and displacement, I attend to three sonic surrogates used in the policing of reproductive respectability. I first examine the revanchist logics that underpin classical music's use on public address systems at Tyne and Wear metro stations in North East England, situating this practice in relation to New Labour's "tough on crime" agenda and the drive for "urban regeneration." Second, I address the adaptation of revanchist logics for urban space in São Paulo, exemplified by the deployment of sound bombs by the Military Police of São Paulo against the expressions of musical and social life that take the form of *pancadões* in the city's poor, Black districts. Third, I draw upon theorizations of postrevanchism, whereby punitive measures are combined or replaced with an ostensibly more compassionate approach to the problems of city life, to consider the Northern Territory Anti-Social Behaviour App, which, amongst other transgressions, enables users to report those creating noise. In so doing, I demonstrate how sound, music, and their associated technologies have been differentially called upon in the management of surplus urban life.

Sleep is an implicit theme that runs through all the book's chapters. It is, after all, integral to processes and practices of reproduction: periods of rest are required for the worker to work, for the child to develop, and for the population to remain mentally and physically well. Consequently, chapter 5, "Lull, Mask, Calm, Renew: Music as Sleep Aid," turns to consider the ways that sound, music, and their associated technologies have been put to work in facilitating, aiding, and enhancing sleep. Beginning with the shared musical resource of the lullaby, this chapter traces the emergence and development of the commercial sound-based sleep aid, first as custom speaker system and second as mobile application. Targeting both the "productive" worker and the household's sleepless child, sound-based sleep aids construe sleep as both a natural and universal desire and a cultivated practice that can be improved through the application of scientific expertise and technological intervention. And if sleep promises to rejuvenate the worker and the household, it also promises to rejuvenate the music industry. Focusing on the sleep and meditation app Calm, and its partnership with Universal Music Group, I turn my attention to the recording industry's convergence with the health tech sector and its implications for music's production. Noting the alignment of functional music with creative freedom, and situating the platform in relation to debates around music streaming and monetization, Calm, I suggest, both confirms and contradicts previous analyses of music's role within the digital economy.

In affirming and enforcing the gendered, economic, and spatial-temporal ideals of reproductive respectability and responsibility, reproductive sound technologies foreclose other possibilities for how life might be made and maintained. Yet it would be a mistake to assume that music only reproduces these norms. The epilogue, "Refusing Reproduction," takes up the refusal manifest in capitalist society's declining birth rates to gesture to some of the ways that music generates friction with capitalism's privatization of need. While I caution against desiring in music the automation of struggle, I suggest that its antagonistic resonances should not be too quickly discounted either. In sounding out the dissatisfactions of capitalist reproduction, music may point to alternative horizons of care and fulfillment.

# 1 Mothering Machines

## PHONOGRAPHY, MECHANIZATION, HOUSEWIFIZATION

On 15 September 1930, daily regional newspapers across the US ran an advert by the American Federation of Musicians seeking signatures in support of the Music Defence League. Entitled "The Robot's Lullaby," the advert contains a cartoon of an automaton dressed in an apron and bonnet, holding a dollar sign rattle (figure 1). Tagged with the label "Canned Music in Theatres," a robotic nursemaid is shown singing to and rocking the cradle of a crying baby, labeled "the audience of the future." Underneath the image, the accompanying text provides a warning to "music lovers":

> "And what becomes of the 'Songs My Mother Taught me'?" murmurs the infant, stirring from a nightmare. "I don't like this new nurse at all, and I want my really truly mother back again." Shall the soothing and refreshing inspiration of Living Music be superseded in our Theatres by the shoddy substitute of mechanical reproductions? Music Lovers by the millions are saying "No!" to this preposterous imposition. The Music Defence League has grown amazingly during the past few weeks. Now it's your turn to join your neighbours in rescuing the sacred Art of Music from machine-made debasement by signing this coupon TODAY![1]

In attaching the threat posed by mechanical music to the nursing of infants, "The Robot's Lullaby" echoes US composer and bandleader John

*Figure 1.* The American Federation of Musicians. "The Robot's Lullaby." *Madison Capital Times*, 15 September 1930.

Philip Sousa's 1906 essay "The Menace of Mechanical Music," which laments the imminent decline of domestic music making and amateur performance due to the introduction of sound recording and player piano rolls. Predicting that children would become indifferent to practice since "automatic music devices" allow music to be heard in the home without "the labor of study and close application," Sousa then turns his attention to the prospective impact of sound's mechanical reproducibility on maternal care: "When a mother can turn on the phonograph with the same ease that she applies to the electric light, will she croon her baby to slumber with sweet lullabys, or will the infant be put to sleep by machinery?"[2] Accompanying Sousa's essay is an illustration by F. Strothmann of an infant in their cradle next to an oversized phonograph horn, being rocked by an "electric oscillator" (figure 2).

Playing on antimodernist sentiments regarding technology and urbanization, which figured the family and the home as a refuge from mechanization,[3] the dystopian images of maternal care in "The Menace of Mechanical Music" and later "The Robot's Lullaby" pay testament to longstanding anxieties about the automation of musical labor. They imply that music, like maternalism, is a natural human resource, and so sound reproduction technologies are a poor surrogate for living, breathing music and its living, breathing musicians. Technological surrogacy for either mother or music is a nightmarish prospect.

"*Will the infant be put to sleep by machinery?*"

*Figure 2.* F. Strothmann illustration, "Will the infant be put to sleep by machinery?," from "The Menace of Mechanical Music." *Appleton's Magazine*, September 1906.

Underlying this technophobic argument is an economic one. While both music and maternal care have frequently been figured as non-work—as activities undertaken due to love or natural proclivity—"The Robot's Lullaby" and "The Menace of Mechanical Music" are underpinned by concerns about the remuneration of performers and composers respectively. "The Robot's Lullaby" reflects the musician's labor union changing attitudes toward music's mechanization. The leadership of the Federation of American Musicians had previously been confident that "machinery" could never replace their work: in 1926, the federation's president

James N. Weber replied to sceptics concerned that radio would eventually threaten job opportunities, stating that "no transmitted musical service will everlastingly displace the desire of the public for personal services rendered by the artist in the presence of the public."[4] By 1930, however, the attitude of the union towards sound reproduction technologies had changed. The economic downturn of the Great Depression had taken hold; US theaters had begun replacing live performers with recordings; the jukebox industry was emerging, displacing musicians in cafes and clubs; and in radio, local bands and orchestras were being replaced with cheaper and often superior music transmitted from powerful and highly capitalized radio stations. The result of these changes was a plunge in musical employment.[5]

While "The Robot's Lullaby" concerns the mechanization of professional musical labor, "The Menace of Mechanical Music" concerns the economic implications of sound reproduction in relation to copyright legislation. Sousa's earnings had been primarily procured from sheet music sales, which relied on amateur musicianship, and from live concerts as a performer. He perceived the new player piano and talking machine industries as a threat to these revenue streams since copyright law offered little recourse for composers seeking compensation for recordings of their music. In the final section of his essay, he offers "a word on a detail of personal interest." Making apparent the tensions between the financial interests of the inventors of the "reproducing apparatus"—who were free to hire soloists and instrumentalists to perform any piece of music in print without remunerating the composer—and those of "the American composer" who earned royalties through sheet music sales, Sousa suggests that "the composers of the music now produced so widely by the mechanical players of every sort draw no profit from it whatsoever."[6]

"The Robot's Lullaby" and "The Menace of Mechanical Music" thus give voice to an apparent crisis in the music industry, whereby sound reproduction technologies disrupt established structures of work and remuneration. Yet implicated in their juxtaposition of the music-reproducing machines with maternal care is an alternative yet related transformation. The presentation of the singing robot nursemaid and the giant phonograph next to the crib as a surrogate for a maternal caregiver, the "really truly mother" as "The Robot's Lullaby" puts it, is a critique the mechanization *of* music but also alludes to music *as mechanization*: its ability to take some of mother's work, as well as that of musicians and composers.

As a means of further explicating these images of sonic surrogacy, this chapter pursues a particular historical narrative concerning the "housewifization" of the phonograph. It focuses primarily on the first two and a half decades of the twentieth century, during which the phonograph became a mass produced and increasingly affordable consumer technology. The phonograph often appears as the first technology to successfully establish sound's recording and reproducibility, transforming how music was made, listened to, and circulated. This attribution is problematic: many elements of what came to be recognized as phonograph culture could be applied to other contemporaneous technologies such as the telephone and player piano.[7] Indeed, to even speak of *the phonograph* in the singular belies a complex and heterogeneous history, and one that necessarily exceeds the "vision" of its paternalistic "papa": the phonograph is best understood as a class of sound recording and playback technologies that have had various uses, markets, technical specifications, materialities, and cultures.[8] Nevertheless, the phonograph is significant due to its proximity to the housewife and her domestic labors: it demonstrates that *sound reproduction* technologies, from the offset, have been *reproductive sound* technologies. To recognize the phonograph as such, I contend, destabilizes its status as a technology of leisure and entertainment. Yet an alternative presentation of the phonograph as a labor-saving technology also requires qualification, insofar as the housewife and the waged worker are distinguished spatially *and* temporally.

Music's treatment as a resource for social reproduction long precedes the phonograph.[9] However, I contend that the mechanical reproducibility of music enables a transformation of this relationship. Two key, related aspects of this transformation can be summarized as follows. Firstly, whereas music was once a resource produced predominantly by the household or community, music's reproducibility enables it to become commodified as a sonic good. As the oft-cited narrative goes, music goes from being something made to something purchased and consumed.[10] This shift can be understood as part of wider transformations within the domestic sphere, whereby money became the primary means of procuring household resources. Secondly, recorded and reproducible music enables reproductive activities to be mechanized: like other sound reproduction technologies at the time, the phonograph is figured as a means of enhancing, assisting or replacing the domestic duties of the housewife. The construction of the phonograph as a tool of household assistance is partly a product of advertising but is also expressive of the gendered division of musical labor and shifting conceptions of the domestic sphere, enabled in part by colonial exploitation and extraction. Consequently, the housewifization of the phonograph exposes the imbrication of gender, technology, coloniality, and the social and economic forms of industrial capitalism in the late nineteenth and early twentieth century.

## SOUND REPRODUCTION, HOUSEWIFIZATION, AND THE TWO SPHERES

"She will in no sense be a drudge, but will be a household engineer. . . . She will finally be able to 'think straight,' will achieve new capabilities, will take the place selfish man has for centuries denied her, and will become the mother of a race of mental proteges."[11] So prophesizes "the phonograph's papa" Thomas Edison, in an 1912 interview for the US women's publication *Good Housekeeping*. He continues: "the housewife will be neither a slave to servants nor herself a drudge. She will give less attention to the home, because the home will need less; she will be rather a domestic engineer than a domestic labourer."[12] This transition—from laborer to engineer, from drudge to intellectual—will be enabled by new technological developments. Electricity, "the greatest of all handmaidens," will do "practically all the manual work about the home," while "other mechanical forces will so revolutionize the woman's world that a large portion of the aggregate of woman's energy will be conserved for use in broader more constructive fields."[13] These technologies are themselves borne of women's dissatisfaction with her domestic duties: Edison declares a general neglect of women's work, with many mothers failing or refusing to "rightly teach their daughters." This situation, however, has advantages: it has "turned the minds of the inventors toward the creation of mechanical devices to perform that work which woman used to do."[14] Freeing up women's time through mechanization will enable progression of the gender and, by extension, of the race: where the woman of the past had an "idle brain," the reduction of her domestic duties will enable her to exercise her intellectual capacities so that she is able to become man's equal, enabling "matrimony to become the perfect partnership."[15] To conclude, Edison contrasts the threat of overpopulation—"the dire fact that this world is getting settled at a rate that will presently occupy its total space"—with an alternative reproductive future, where less space is occupied by the "unfit and the imperfect." In this future, where she has been liberated from domestic labor, woman becomes "the wife, not a dependent; not alone a mother, but the teacher and developer," producing a new generation of moral, mental, and physical prodigies.

Edison's vision for women's technological liberation from domestic labor reflects the growing availability of electricity and associated consumer technologies such as the vacuum cleaner and the electric stove. Yet he also draws attention to the phonograph, pointing to a future where "the drudgery of life will, by and by, entirely disappear" and "both men and women will be gratified by an infinite variety of delightful sights, sounds, and experiences that today are unknown and unimagined." In a short time, Edison suggests, it will be possible for the humblest home to access the music of the world. Indeed, the mechanical reproduction of music will provide a superior listening experience to the live event: unlike the home listener, the listener in the auditorium may hear too much brass or strings, depending on the placement of their seat. Developments in sensory media—the phonograph but also the "home picture machine"—will "do more for the development of women than they will for the development of men," helping to grow their cognitive abilities.[16]

Titled "The Woman of the Future," Edison's striking interview is expressive of ideological, social, and technological transformations occurring within the domestic sphere in the early twentieth-century, as the Victorian ideal of the "angel of the home"—an ideal that many women lacked the material means to approximate—was reconfigured by a "Taylorism of the home," aided by the proliferation of new appliances, technologies, and instruments.[17] Here, the replacement of mother with machine is not a nightmarish prospect but one that carries multiple benefits, from improving the qualitative character of housework (from manual drudgery to intellectual pursuit) to mitigating overpopulation and ensuring the enhancement of the human race. Yet the "dependent" woman the inventor seeks to liberate is herself an industrial invention; a correlate of the spatial and temporal separation of production and reproduction that Maria Mies refers to as the "housewifization" of women in the colonial core.

Housewifization emerges with the intensified distinction between two spheres of social and economic activity: the public sphere of waged work and the private sphere of domestic labor. The latter represents the externalization of costs that would otherwise need to be covered by capitalists, inasmuch as women's domestic labor comes to be considered a natural resource, "freely available like air and water."[18] Indeed, housewifization is predicated on the devaluation, naturalization, and feminization of socially reproductive activities—of "mothering," caring, and maintaining the home—as households become increasingly reliant on the wage. For Mies, the housewifization of women and their consignment to the domestic sphere of social reproduction is constituted by two stages.

The first stage of housewifization is predicated on the creation of women in "accumulating classes" as consumers and demonstrators of luxury and wealth. Fueled by colonial extraction, the market for luxury items expanded over the course of the eighteenth and nineteenth century with the imitation of the European courts by the *nouveaux riches*.[19] With this expansion, certain trends in luxury production emerged, including a tendency toward domesticity and privatization. Mies proposes that the alignment of the luxury market with domestic space influenced the image of the "good woman" in capitalism's centers in the nineteenth and twentieth centuries, "namely woman as *mother* and *housewife*, and the family as her arena, the privatized arena of consumption and 'love,' excluded and sheltered from the arena of production and accumulation, where men reign."[20]

The second stage of housewifization concerns the establishment of the nuclear family as social, economic, and sexual norm, consisting of husband as breadwinner, nonearning legal wife, and children. In the eighteenth and early nineteenth centuries, marriage and the family form—predicated upon the combined principles of kinship and cohabitation, and under protection of the state—were possible only among the propertied classes of the bourgeoisie: "It was the bourgeoisie which established the social and sexual division of labour, characteristic to capitalism. The bourgeoisie declared 'family' a private territory in contrast to the 'public' sphere of capitalism."[21] The poor and propertyless were exempt from the family form's heterosexual division of labor: "amongst peasants, artisans and workers, women had always to share all the work."[22]

In the second half of the nineteenth century, a series of European legal reforms expanded the bourgeois family form among the proletariat. Prior to these reforms, many of the industrial proletariat had been women and children. But this had deleterious implications for the reproduction of the workforce, given the impact of overwork and working conditions on women's health: Mies contends that these women had "no particular material interest in producing the next generation of miserable workers for the factories. But they constituted a threat to bourgeois morality. . . . The proletarian woman had to be housewifized too, in spite of the fact she could not afford to sit around and wait for the husband to feed her and her children."[23] Church and state directed legal and moral prohibitions against abortion and sexual activity outside of marriage, serving to police the family as the primary institution through which children were produced. In the US, housewifization was accompanied by violent campaigns against Indigenous communities, Mormons, and other social groups who were not exclusively reliant on capitalist relations and whose forms of life did not conform with the bourgeois family form.[24] Yet among the "the most advanced sections" of the working class, primarily, "the male skilled labour aristocracy in European countries" there was support for the production of the bourgeois nuclear family among the working class: Mies proposes that while working-class women had little to gain from housewifization, for working-class men it meant the monopolization of waged work and financial control. Thus, contra Marx and Engels, Mies demonstrates that capitalism did not destroy the family but created it: first among the propertied classes, then the working classes. With it came the category of the housewife.

As her role changes from laborer to engineer, from drudge to operator, Edison's woman of the future remains responsible for the domestic sphere, even if its demands on her time are reduced and the conditions of her labor improved. Indeed, as the nineteenth-century "angel of the home" became the twentieth century's "Taylorist of the home," the separation of the two spheres was maintained. The emergence of new consumer technologies, meanwhile, affirmed the home as a key site of capitalist consumption. While pre- and early capitalist households acquired resources of life making and maintenance through a range of means—direct production, inheritance, gifts, or barter, for example—money became the primary means of procuring household resources, which were purchased as commodities, over the course of the nineteenth century.[25] In the early twentieth century, as the household became an important market for new technologies and mass-produced commodities, Mies suggests that the accompanying "new ideology" of scientific home management served to facilitate the further domestication of women: "Not only was the housewife called on to reduce the labour power costs, she was also mobilized to use her energies to create new needs."[26]

The scenes of sonic surrogacy depicted in "The Robot's Lullaby" and "The Menace of Mechanical Music" attest to the "housewifization" of sound reproduction. As manufacturers sought to communicate what these technologies were for, women in the home were regularly presented as their primary beneficiaries: the phonograph, like other early sound reproduction technologies, was mediated by bourgeois configurations of the domestic sphere as a private site of renewal, reproduction, and consumption. It was women's domestic and musical labor that helped determine the function and purpose of the phonograph, and the associated successes of other household musical commodities: the phonograph was preceded by and for some years existed alongside the sheet music and upright piano industries, both of which recognized women as key producers and arbiters of music in the home.[27] Thus, to cite a 1908 National Phonograph Co. advertising campaign, "every woman, every mistress of a home and every hostess" should have a phonograph (figure 3). Women and children were frequently represented as the phonograph's operators and listeners. Adverts touting the convenience and versatility of the phonograph, and its ability to bring culture and class into the home were regularly featured in women's magazines. And women came to be recognized as the primary consumers of the phonograph by manufacturers: a 1919 study by the Sonora Phonograph Company showed that women purchased players more than twice as often as men. When men accompanied them into stores, women remained the primary decision-makers.[28] After all, "phonographs were just domestic consumer products like vacuum cleaners and refrigerators."[29]

*Figure 3.* "Every mother, every mistress of a home, every hostess should have an Edison phonograph," from *Ladies Home Journal*, 1908.

The housewifization of the phonograph, however, was neither complete nor immediate. In its nascent years, Edison proposed many possible uses for his new technology, which notably included the recording and reproduction of lullabies. Foreshadowing Sousa's anxious polemic, he writes in his 1878 *North American Review* article "The phonograph and its future" that the technology will "allow the child to form their first songs, or sing them to sleep."[30] Nevertheless, the early phonograph was understood foremost as a technology for the workplace, envisaged and marketed as a voice-recording apparatus for the office, enabling businessmen to "*dispense with the clerk*, and *maintain perfect privacy* in their communications." As well as mechanizing the work of a stenographer, Edison understood the phonograph to increase the efficiency of communication, proposing that the "talking machine" would enable the recipient to listen to his letters "being read at a rate of from 150 to 200 words per minute, and at the same time busy himself with other matters."[31] Edison's persistent emphasis on the recording capabilities of the phonograph, however, meant that it struggled to find a commercial market.[32] The domestic potential of the phonograph only began to be realized in the early 1890s as it became legible as a musical instrument for the home.

In keeping with Victorian values, the early domestic phonograph—at this time a luxury good only available to wealthier families—promised to provide household with morally and aesthetically virtuous entertainment and education: it offered the opportunity to own the voices of "great" singers and speakers, which could be played on demand or assist in vocal training.[33] Yet the phonograph's status as a "respectable" addition to the home was troubled by its use as a source of cheap, popular amusement within arcades, parlours, and transportation hubs.[34] Introduced in late 1889, the coin-operated automatic phonograph played short recordings—often of popular vaudeville and minstrel songs—and was the means through which many US listeners first encountered the technology. While the success of the coin-operated phonograph played an important role in establishing the reproduction of music as the phonograph's primary purpose—its association, in its early years, with cheap entertainment and urban working-class neighborhoods came to be eagerly disavowed by the phonograph companies as they pursued middle-class consumers.[35]

While phonograph companies were keen to emphasize its compatibility with domestic life, sound reproduction technologies troubled established bourgeois norms of the home by rendering its boundaries porous. The telephone was "the first electric medium to enter the home and unsettle the customary ways of dividing the private person and family from the more public setting of the community," providing women with new opportunities for social contact, and enabling "the office" to be accessed from home.[36] There were similar tensions between the phonograph and the Victorian conceptions of domesticity that its own marketing and circulation referenced: Holly Kruse argues that, as one of the first mass-produced, technologically advanced entertainments to enter the home, it faced an unprecedented struggle inasmuch as the middle-class home of the late 1800s was imagined as a retreat from modernity and a hostile "outside," while mass entertainment was meant to be a public rather than private event.[37] As Carolyn Marvin points out with reference to the 1890s gramophone, "Media were to adorn not transform home life. Home was still a quiet place of culture, where the routines of women would go on supporting men's forays into less protected places."[38] Consequently, phonograph companies sought to minimize any perceived threat sound reproduction might pose to the privacy and propriety of the home, figuring it as an edifying technology of education and enlightenment.

Sound reproduction technologies emerged at a time when both gender roles and the domestic sphere were being gradually reconfigured with and alongside new consumer technologies and mass production. But their apparent capacity to traverse, as well as affirm, the two spheres serves as an important reminder that housewifization and the gendered division of labor was never complete nor uncontested. For poor, racialized, and working-class women, the ideals of domestic femininity were impossible to satisfy. Many women undertook waged work before marriage, while non-white and immigrant women were likely to continue to work as married women due in part to racially segmented labor markets. Often, women's waged work was clustered around activities that closely resembled those of the home: "the textile and garment industries, laundering, nursing, service work, and especially domestic work in someone else's home."[39] The phonograph itself was associated with gendered employment opportunities. Music recording saw a minority of women developing and furthering their careers as performers, particularly as vocal performers: indeed, as Lisa Gitelman elucidates, women's singing voices, which were figured as particularly challenging to record well, became a kind of standard, testifying to the "realist" reproductions of the phonograph and the technical superiority of one product over another.[40] Furthermore, selling, as well as buying and using the phonograph, became women's work. In the final decades of the nineteenth century, the number of women within the labor force of department stores rapidly expanded, and young women were increasingly involved in the retail of phonographs and records at the turn of the century.[41] The housewifization of women's labor, then, was by no means a totalizing and immutable process, nor was it synonymous with social passivity: women's organization and participation in consumer boycotts and cost-of-living protests, including in the first decades of the twentieth century, make clear that the "separate" sphere produced by housewifization has been a terrain of struggle.[42] Nevertheless, the feminization of household reproductive activities and associated practices of consumption underpins sound reproduction technologies, their gendered imaginaries, and social practices.

## MECHANIZING THE PIANO GIRL

A baby with an excited expression sits on the knee of an older woman, hands apart and arms stretched out (figure 4). The infant is not reaching for its mother, who stands behind the desired object, but an Edison phonograph, whose horn is turned toward the delighted young listener. Appearing under the tagline "Everybody's happy when the phonograph plays," this scene of domestic enjoyment illustrates the device's utility: "For the baby a lasting diversion—for grown-ups a pleasure that never loses its novelty."[43]

Thomas A. Edison Inc.'s 1911 advert once again situates sound reproduction amid the maternal-infant dyad. Here, however, the machine does not replace the maternal figure. Rather, foreshadowing Edison's descriptor of the housewife as domestic engineer, she appears as the phonograph's operator, working *with* the "instrument" to facilitate the musical pleasure of the infant and the accompanying "grown-ups." If "everybody is happy when

*Figure 4.* "Everybody's happy when the phonograph plays," from *Success Magazine* & the *National Post*, 1911.

the phonograph plays," then we might infer that her happiness partly stems from its ability to divert, distract, and entertain on her behalf.

As a home technology, the phonograph has been frequently framed as a means of entertainment, providing respite and relief from both waged work and domestic drudgery as it became an increasingly affordable commodity. Noting the historic coherence between the emergence and popularization of sound reproduction technologies on the one hand, and the emergence of Taylorism and Fordism on the other, Timothy Taylor notes how the "banalization" of work heightened the importance of consumption as a leisure activity as a means of escape.[44] Indeed, the phonograph is emblematic of a stricter separation of work and leisure: sound recording and reproduction further concretize music as something that occurs outside of the times and spaces of waged labor, be it in the home or other sites of consumption such as theaters, clubs, and music halls.[45] For women within the domestic sphere, the sound reproduction technologies with which they were representationally allied ameliorated the emotional and physical effects of their domestic responsibilities: whereas for the waged worker, the phonograph offered an escape from the demands of labor from within the comfort of their own home, for the housewife the phonograph promised an escape *from* home *within* the home. As William Kenney notes, the phonograph was said to "offer solace and 'transcendence' to the overworked and lonely housewife and a burgeoning, exciting and swiftly evolving world of recorded sounds into which women were encouraged to retreat."[46] Thus, in providing pleasure, entertainment, and leisure, sound reproduction technologies could soften the effects of monotonous work across the gendered division of labor.

To consider the role of the phonograph solely in terms of entertainment and escapism, however, risks obscuring the ways that sound reproduction technologies enacted more direct interventions into the production and maintenance of home and family. The proximity of the phonograph to other household consumer appliances such as vacuum cleaners is instructive, as is the recurrent trope of the infant listener: the phonograph interacted with women's labor as homemakers and carers. Frequently framed as a domestic aid, the phonograph was positioned as assisting in and fulfilling demands for the provision of musical listening and "rational amusement"; the affective management of the household; the maintenance of familial ties; and the education, entertainment, and "diversion" of children. In this regard, the phonograph belongs to an expanding field of consumer technologies mechanizing women's work within the home.

Such a claim may appear counterintuitive, for it rubs against common conceptualizations of both labor and music. It might be contested, for example, that the comparison of the phonograph to the vacuum cleaner—or indeed the iron or electric stove—is rather limited: where the latter can be easily framed as labor-saving technologies in that they transform household chores such as cooking and cleaning, the instrumentality of the phonograph is less immediate. Unlike the vacuum cleaner, sound technologies were not explicitly framed as reducing or replacing the work of the housewife or the work of the servant.[47] Indeed, Edison's interview on the woman of the future appears to hold the phonograph apart from other consumer technologies available to women in the home; it provides cognitively beneficial sensory stimulation rather than mechanizing domestic drudgery.

The rationale for understanding the phonograph as labor-saving technology becomes clearer when we consider who made music in the home prior to its mechanical reproducibility. Like other life-making and life-maintaining resources such as clothing and food, music was procured and produced in various ways prior to its reproducibility and commodification. However, in the bourgeois household, domestic music making was an activity primarily associated with daughters—but also with wives. For Ruth Solie, this gendered division of musical labor was informed by a "vulgarized form of romanticism that idealized and sentimentalized women as it idealized and sentimentalized aesthetic experience, creating a natural link between them."[48] Among the nascent middle classes of the nineteenth century, the ability to play the piano indicated women's proximity to genteel society, and learning to play was a standard aspect of a girl's training: it was believed to provide her with "discipline, diversion and a skill that would help her attract a husband."[49] Women's piano playing was both an expression of leisure and a form of moral and emotional labor within the family. The piano and its usage symbolized of a life free from toil, allowing for the development of both moral and musical refinement. The private refuge of domesticity with which the piano was associated was something that could be maintained through the patriarch's success in the public world of business and industry.[50]

The leisure of the piano also informs ideals of performance practice: in contrast to the "confidence" of the professional musician, feminized amateur music in the parlor was to remain "modest" in its articulation. However, like the professional musician, the parlor pianist was to appear "effortless."[51] This effortless performance was the result of hard work: indeed, from the mid-nineteenth century, US piano instructors increasingly insisted that proper practice should be physically demanding, encouraging the development of finger strength and dexterity through hours of intense and painful exercises.[52] Within the well-regulated home, the provision of leisure listening and "rational amusement" for family members was one domestic chore among many. Performance at the piano was one way in which women fulfilled their responsibility for the emotional and moral well-being of the family: music was heard not only as entertainment but was, in Solie's words, "a sort of combination spiritual therapy and mental hygiene."[53] Adolescent daughters, meanwhile, were ideally suited for providing this musical labor, able to "perform this important service role within her birth family while at the same time practicing for her own wifely and motherly career."[54]

For both critics and advertisers, the phonograph came to be viewed as a surrogate for women's music making in the home.[55] Indeed, the "menace" that "mechanical music" portents for Sousa is largely due to the transformation and alleviation of feminized musical labor by these new sound reproduction technologies. Whereas the Victorian wife was expected to soothe her husband with the parlor piano and edify her children with appropriate music, the phonograph provided an alternative—and purportedly more efficient—means of undertaking these duties: as Sousa feared, the phonograph meant that music in the home was no longer reliant on hours of piano practice, nor on the sourcing of sheet music. In separating music in the home from the women who produced it, the phonograph reconfigured, as well as affirmed, music's gendered distinctions: as Mark Katz makes clear, the phonograph enabled men to become listeners and enthusiasts of a music culture that was historically designated as feminine, since the proximity of the "machine" mitigated against music's "unmanly" influence.[56] It would be inaccurate to suggest, however, that the phonograph simply and rapidly replaced the piano-playing woman: as Dave Laing notes, they existed alongside one another for a time, with the phonograph only beginning to displace the piano in the 1910s.[57] And while the phonograph certainly garnered attention and enthusiasm from men in the home, it was women who were typically figured as the phonograph's *users*.

Marketing materials for the phonograph offered the housewife assistance in entertaining and educating the various members of the household: in a *Ladies Home Journal* from February 1910, an advert for Edison's Amberola is accompanied with an offer of "a book for every woman free." Listed alongside sheet music for a waltz song, readers are offered a copy of "The Edison Phonograph and the Home," which details the moral imperatives of music for the family. The phonograph is positioned as alleviating women's "responsibility" for "amusing" a husband and children of different ages: "If you have a home full of lively youngsters with their friends, some grown-up children and a husband for whose amusement you feel a certain responsibility, then we have a book intended for you particularly."[58]

The 1910 Amberola advertisement suggests that, as with the piano-playing girl before her, the musical choices of the phonograph's housewife-operator were not only aesthetic but functional. They were to aid with the fulfilment of domestic responsibilities, determined in accordance with bourgeois norms of reproductive respectability. While the advent of sound recording has often been understood as separating music from social function (for example, music originally made for dancing in communal spaces, once recorded, becomes "autonomous" music for consumption in any time or place), the housewifization of the domestic phonograph complicates this narrative, since recorded music's commodification and reproducibility enabled it to take on different functions within the home.[59] These functions include the affective management of the household. Where women's music in the Victorian home served to establish appropriate moods and help secure the emotional well-being of the family, reproduced music was similarly deployed by wives and mothers as a means of regulating emotion in domestic space.

Such uses of recorded music as an affective technology find articulation in Thomas A. Edison Inc.'s *Mood Music* program, which began in 1920 under another suggestive title—*What Music Will Do For You*—and was overseen by William Maxwell, the head of the Musical Phonograph Division. The program represented an effort to formalize music's effects on body and mind for the purpose of categorizing, collating, and consuming recorded music. Part research, part marketing campaign, *Mood Music/What Music Will Do for You* was itself a product of a collaboration between the Edison company and a group of psychologists at the Carnegie Institute of Technology, including the renowned industrial psychologist Walter V. Bingham, which sought to explore the psychological effects of music. The classification scheme that emerged from the program consisted of twelve distinctive moods, which in the promotional *Mood Music* booklet are described as covering "almost every need for daily life." Categories included "For More Energy!," "To Bring You Peace of Mind," "Stirring," and "For the Children," with each accompanied by a list of recommendations from the Edison Re-Creation record catalogue. Accompanying illustrations demonstrated music's ability to refresh and reinvigorate men and women exhausted from work or running errands. *Mood Music* thus sought to figure recorded music as something practical and useful, enabling the listener to make musical selections for the purposes of self-improvement.[60]

*Mood Music* primarily portrays musical listening as a private, individual, and domestic experience: listeners (both male and female) are positioned as using the phonograph to alleviate or modify their own affective states.[61] In this, it preempts one of the prominent modes of musical curation and consumption in the twenty-first century's streaming economy: the notion that recorded music can be collated and consumed as a means of modulating a listener's physical and psychological state has become a familiar truism. Yet, as will become a recurring theme over the following chapters, music's presentation as a technology of the self is shadowed by another use: music as a tool for managing the nuclear family. This is represented here by the final mood category, "For the Children," which by its very presence implies a separation between those *using* music within the household (i.e., the "mothers and mistresses of the home" referred to in figure 3) and those *listening to or entertained* by music (i.e., the children). This differentiation between user and listener is similarly traceable in the domestic scene presented in "Everybody's happy when the phonograph plays" (figure 4), where the phonograph's user (seemingly the infant's mother) is distinguished from the listener (the infant, alongside other "grown-ups"). Although this distinction remains necessarily leaky—the phonograph's operator is also listening to music, even as she is using it to "divert" her infant—this repeated separation of the phonograph's operator and listener(s) troubles the social and musical relations often assumed by discussions of recorded music's consumption and experience, whereby listener, consumer, and user are often assumed to be one and the same.

In 1921, The Edison Company distributed copies of a questionnaire asking phonograph owners in forty-three states about their favorite. Connected to Bingham's work into the psychological effects on music, the company reported to have sent out 20,000 copies of the survey. As William Kenney highlights, the responses contain narrative comments documenting a range of ways in which the phonograph and its records were engaged with in the home and a variety of popular attitudes towards the technology. Repeated themes include sound recording's relationship to family identity; its capacity to evoke memories of home, childhood, and loved ones; its ability to facilitate grief; and its use as a mediator between immigrant and US national identity. However, the survey also reveals ways in which recorded music was used by women to manage interfamilial relations and collective mood within domestic space. Men and women answered the survey in roughly equal numbers but, among answers by women, a figuration of the phonograph as a household tool emerges. As Kenney notes, "many female respondents revealed that they strengthened family ties in the tired irritable hours in the late afternoon when relations became brittle."[62] One mother, for example, suggests that "they have all moods of music. Some one looks *blue* or a little peeved [and] I go in and start 'Henry Jones Your Honeymoon Is Over' and at once everybody smiles and the white flag waves."[63] Another suggests she uses the phonograph to keep her family happy. Consequently, the phonograph had become an indispensable necessity for the household: "If we had to part with a piece of furniture in our home, we would give up our bed before we would give up our Edison."[64]

As well as helping with the social and affective management of the nuclear family, music reproduction technologies promised women assistance in maintaining social bonds with household visitors. A 1913 advert by Victor—a key competitor of the Edison Company during the first decades of the phonograph industry—is instructive in this regard. Unlike Edison's machines, which used cylinder records Victor's device used flat discs derived from Emilie Berliner's gramophone: a format that would eventually become the industry standard. Like other phonograph companies, Victor had sought to disavow the technology's early associations with vaudeville, the arcade, and cheap entertainment. It did this by engineering improvements to audio fidelity, expanding its operatic repertoire and transforming the utilitarian machine of sound reproduction into a luxury furniture item for the home.[65] The company's construction of a "high class" status for its devices is traceable in their adverts proclaiming "The best friend of a hostess is a Victrola."[66] Depicting a lavish room filled with well-dressed guests, it assures its audience, "The hostess who has a Victrola never need worry about how the evening will 'go.'" The product in question is set at the center of the room next to a hosting couple, the hostess greeting a guest while others look towards her. The machine, described as "never obtrusive but always ready," is said to "break the ice" and prevent awkward silences and to entertain young people with the latest dance music when they tire of general conversation. The purchase such a device thus enables the woman of the home to transfer a degree of responsibility for entertaining and engaging guests onto the machine.

The housewifization of sound reproduction thus enacted changes to the sphere of social reproduction by mechanizing aspects of its provision. The phonograph and its music acts as a surrogate for the piano-playing girl of the bourgeois Victorian home, mechanizing women's musical labor and enabling recorded and reproducible music to take on new use-values relating to the social and affective management of the home. While the musical labor of the Victorian piano girl itself performed ideological, affective, and social functions in alignment with by ideals of gendered propriety and respectability, the domestic reproducibility of music was accompanied by a reassertion of its domestic duties: from diverting infants to breaking the ice at social functions. In this regard, mechanized music itself becomes a kind of mechanization, functioning as a household tool used by wives and mothers to assist, enhance, and alleviate their work within the home.

## TIME, MEASURABILITY AND THE GENDERED DIVISION OF LEISURE

If the machines of the late nineteenth and early twentieth century could reduce the demands of home and household, then it might be thought to follow that they expanded women's leisure time: allowing "father" to "do his own playing," to paraphrase an 1913 article from the journal, *Musician*, would allow women partake in other activities.[67] In his compelling study of gender and the phonograph, Nathan Bowers exemplifies this line of reasoning: Bowers suggests that in alleviating the need for domestic musical labor, the phonograph enabled women to enjoy music recreationally: "Eliminating the need of a pianist freed all members of the household to participate in recreational activities together."[68] This supposition was reproduced in advertisements, with the phonograph positioned as a means of providing women listeners with "brightness and cheer," enabling them

*Figure 5.* "The rivals," from *Munsey's Magazine* 1909.

to put "out of mind the little trials and worries of the day" in the evening, once work is done and children are in bed.[69] The recurrent figure of the white woman listener in visual depictions of sound technologies similarly affirms the association of the phonograph with her domestic recreation (figure 5).[70] Yet just as the effortlessness of the piano girl should not be mistaken for the absence of labor, these claims to freedom afforded by sound technologies should not be accepted at face value.

The time of leisure, which sound reproduction technologies are credited as creating and filling, is dictated by and separate from the time of work. Indeed, the compartmentalization of work time and leisure time is a historical correlate of housewifization, and the separation of production and reproduction into two distinct spheres of activity. Whereas labor and life were previously intertwined, the establishment and intensification of what E. P. Thompson calls "clock time"—driven in part by industrial capitalism's need for synchronized labor (see chapter 5)—meant that the times and spaces of work, rest, and leisure became disentangled and formalized.[71] Thus, as the historian Peter Bailey notes, "in the populous and extensive industrial city, leisure was time clearly marked off from work, to be pursued elsewhere than in the workplace and its environs, and undertaken in company no longer in the nature of things comprised predominantly of workmates."[72] Furthermore, work and leisure were no longer regulated by the customs and obligations of community life, meaning that the characteristic recreations of the second half of the nineteenth century were substantively different from their predecessors: "not only were they in many cases the products of a new technology and new social groupings, but they now took place within the unique circumstances of modern leisure, a condition of individual free choice specific to industrial society and a qualitatively new dimension in the experience of the masses."[73] Although unevenly distributed, leisure became an area of relative autonomy in the everyday life of all classes; by the 1860s, it was considered "a necessary amenity, a basic overhead in the maintenance of an industrial society."[74]

If leisure is the time that "lies outside the demands of work, direct social obligations and the routine activities of personal and domestic maintenance,"[75] then the amount of this time available (and how it is to be spent) is shaped by gendered expectations, coconstituted by race and class. Indeed, insofar as they assume a clear separation between the times and spaces of work and recreation, analyses of leisure are frequently grounded in the experiences of the white, masculine waged worker. Yet for those whose "free time" outside of work hours is to be filled with domestic duties and caring responsibilities, it can be difficult to distinguish leisure from nonleisure activities. In seeking to trouble the definitions of what constitutes work by recognizing women's reproductive labor as integral to capitalist society, feminists have also troubled what constitutes leisure, highlighting how activities coded as such have been reliant on feminized and naturalized labor.[76] Interrogating the trope of the "idle" nineteenth-century middle-class wife of industrial capitalism, for example, Anne McClintock proposes that leisure constituted a "character role" performed by women seeking membership in the "respectable" classes. Recalling the "effortless" labor of the piano-playing girl, the substantive domestic demands placed upon women in the emerging middle classes was intensified by a need to conceal this labor. "Even with some assistance from an overworked yet invisible maid, 'idleness' was ultimately a performance rather than the absence of obligations, demands and hard work. Housewifery," McClintock states, "became a career in vanishing acts."[77] Others have made apparent how the activities that fill many women's free time are difficult to categorize vis-à-vis the work-leisure distinction. "If you are doing a time study of housewives," asks Ruth Cowan Shwartz, "are you supposed to define the time they spend watching their children play in the park as leisure or as work?"[78] In light of this question we might revisit, once again, the maternal phonograph operator figured in Edison's advert "Everybody's happy when the phonograph plays" (figure 4): in using the phonograph to entertain the infant, is she participating in musical leisure time or undertaking domestic labor?

These gendered limitations of the work-leisure distinction are similarly recognizable from secondary accounts of women's engagements with sound reproduction technologies. In his analysis of the early residential telephone, Claude Fischer suggests that women's adoption and adaption of the telephone to their own ends serves to challenge critiques of domestic technologies that envisage them as increasing women's domestic burden and responsibility. Like the phonograph, the telephone began as a work technology—a way to connect the businessman's home to his office—but, upon entering the domestic sphere, women became informal users of the technology. From this, a telephone culture was established that associated the technology with the housewife and her domestic activities.[79] For Fischer, women's use of the telephone was less to do with homemaking—as might be the case with other mechanized devices such as the sewing machine or washing machine—and more to do with socializing and conversing with family and friends. Among the various explanations offered for the affinity between women and the telephone, Fischer posits that women's established responsibilities for "socioemotional" tasks and the management of social networks contributed to the gendered patterns of the telephone's uptake. However, Fischer suggests that explaining the telephone's adoption by women in terms of the gendered division of labor is incomplete: "it is implied that conversation is more housework, only another burden. It does not capture the degree of voluntary conversation and the pleasure in conversation suggested by the accounts we have of women and telephony." The enjoyment women derived from telephone conversation, in Fischer's view, undermines any attempt to position the telephone as simply intervening in—and adding to—women's housework. Fischer does concede that the telephone, in facilitating greater connectivity, may have made women more vulnerable to calls for socioemotional labor from family and friends: it may have contributed to increased demands for advice, comfort, and organization. But ultimately, women readily used the telephone for "what they *wanted*: sociable conversation."[80]

Fischer's account of the residential telephone aims to trouble what he articulates as a problematic generalization, whereby domestic technologies simply concretize or exacerbate the gendered division of labor. But his partial dismissal of the telephone as enabling and generating housework is revealing of some of the complexities in accounting for reproductive labor vis-à-vis the work-leisure dialectic. Work, in Fischer's account, appears to be defined affectively: activities are work if they feel like a burden. Marx and Engels similarly describe work as a joyless activity, having "lost all individual character, and consequently, all charm for the workman."[81] Yet Marx and Engels make clear that this joylessness is symptomatic of a particular set of waged working conditions under industrial capitalism rather than being inherent to work itself. It is possible for work under capitalism to be perceived of as satisfying or pleasurable without contradicting its exploitative character.[82]

Whether it is concluded that women's uses of the early residential telephone—or indeed the musical activities of the phonograph's housewife-operator—appear as leisure but are actually labor, or that they combine aspects of both, Fischer's account necessitates further consideration of women's domestic and socioemotional activities vis-à-vis the category of work, too. At stake here are tensions between sociological and technical understandings of different types of labor. Indeed, this has been a key point of focus among Marxist feminist analyses of reproductive labor, where there has been much discussion as to whether women's household, domestic, and caring activities should be considered productive. For Marx, productive labor (often taken to be synonymous with work) is not simply labor that creates something—what he refers to at the beginning of the first volume of *Capital* as "useful" labor.[83] Rather, productive labor is a social relation specific to the capitalist mode of production and concerns the generation of surplus value. Leopoldina Fortunati, among others, argues that reproductive labor undertaken by women in the home is in fact productive—or rather, she contests the separation of production and reproduction on the grounds of value creation, noting that "although reproduction appears as the creation of non-value, it clearly contributes to the creation of value as a crucial, integral part of the capitalist cycle. . . .

the real difference between production and reproduction is not that of value/non-value. . . . reproduction is the creation of value but appears otherwise."[84] Referring to housewives and sex workers as partaking in "indirectly waged labour," Fortunati argues that the gendered division of labor sees the male breadwinner exchange his wages for women's labor power. On this basis, Fortunati argues that this constitutes a relation of production: "although this exchange appears formally to take place between the man and the woman, in reality it is the form that the relation of production between women and capital takes, which is a non-directly waged exchange."[85]

Others, contra Fortunati, have sought to maintain the formal separation of production and reproduction: reproductive labor, while necessary to capital and its modes of production, remains largely external to market relations governed by the value form. This line of argument is exemplified by Endnotes Collective, who argue that "there must be an exterior to value in order for value to exist. . . . we would say that, for labour-power to have a value, some of these activities have to be cut off or dissociated from the sphere of value production."[86] Reproductive labor, as that which produces labor-power and maintains those excluded from waged work, can be envisaged as the production of "pure" use-value. In this regard, as Marina Vishmidt and Kerstin Stakemeier propose, it is exceptional in relation to the regime of value.[87]

The debate as to whether reproductive labor is productive of value has itself been critiqued for prioritizing formal concerns over the political questions raised by Fortunati and others regarding capital, gender, and exploitation.[88] In any case, what is of greater significance here is the related but distinct argument concerning the *time* of productive and reproductive labor as it is undertaken by the housewife. In the context of productive labor, value is an expression of labor-time. When labor power is sold, it is done so in relation to a specified amount of time to be worked. The wage, meanwhile, marks an expectation of what can be achieved within an allocated amount of working time, in alignment with a generalized social average. However, insofar as the reproductive labor of the housewife is not directly market-mediated, its relationship with time is different. As Mariarosa Dalla Costa and Selma James suggest, "If you are not paid by the hour, within certain limits, nobody cares how long it takes you to do your work."[89] For Dalla Costa and James, the temporal difference between reproductive domestic labor and other kinds of labor is not only quantitative but also qualitative:

> If technological innovation can lower the limit of necessary work, and if the working-class struggle in industry can use that innovation for gaining free hours, the same cannot be said of housework; to the extent that she must *in isolation* procreate, raise and be responsible for children, a high mechanization of domestic chores doesn't free any time for the woman.[90]

While specific tasks may be made more efficient—cooking, cleaning, or washing—this does not reduce the time of the working day. The housewife, whether or not they are also employed outside the home, "is always on duty."[91]

The measureless quality of a housewife's working day, then, troubles the notion of labor-saving domestic technologies. As Fischer's account of the telephone alludes to, feminists have frequently argued that technological innovations within the domestic sphere have rarely saved women's time. While industrialization and technological advancements promised to reduce the amount of effort and exertion necessary to maintain the home, domestic technologies change the specificities of housework without limiting the expectations placed on wives and mothers.[92] Domestic technologies or "efficient" housekeeping might limit the time spent on particular tasks but the housewife nevertheless remains on duty: as Dalla Costa and James put it, "her workday is unending not because she has no machines, but because she is isolated."[93]

On this basis, the labor-saving appearance of the phonograph and other sound reproduction technologies requires qualification. Even if the phonograph entertains young children, for example, the housewife is likely still required to operate the machine or keep an eye or ear out while she undertakes her other household responsibilities. After all, as Dalla Costa and James imply, young children still require care for a full day and this time is inflexible, even if the time dedicated to specific activities can be reduced through mechanization.[94] This is not to deny that housewives listened to the phonograph recreationally, nor is it to downplay the transformations enacted by the phonograph to music culture within and beyond the home. Nevertheless, the recurrent trope of the leisured woman listener exists in tension with the time difference that characterizes the two spheres.

## TECHNOLOGICAL SURROGACY AND HIERARCHIES OF THE HUMAN

The act of sonic surrogacy portrayed in "The Robot's Lullaby," with which this chapter began (figure 1), is a racialized encounter. The distressed infant is white. Meanwhile, the mechanized carer, whose presence fails to comfort them, is marked as Black: the illustration's shading, combined with the bonnet and apron outfit recalls the mammy stereotype prevalent in US culture. There is thus an implied double replacement of "really truly mother": first by the Black nursemaid, with whom some of the housewife's childrearing and domestic duties are outsourced, and second by the Black nursemaid's automaton.

Whether presented as a domestic aid for the housewife or a dystopian surrogate for motherly love, the positioning of sound reproduction as a technology of mechanization is grounded in gendered *and* racialized figurations of the human-machine relationship. As Neda Atanasoski and Kalindi Vora elucidate, since the first industrial revolution, technological automation has signaled the threat and promise of the replaceability of specific types of human functions and workers. Tasks that are deemed mechanizable, including reproductive and care work, were regarded as noncreative, and consequently could be undertaken by the poor, the uneducated, the colonized, and women. Edison's "woman of the future" is once again exemplary in this regard. It is not only her own domestic drudgery that the machine will replace but, in more affluent homes, the labor performed by servants: "The housewife's work, in days to come, will amount to little more than superintendence, not of Norah, fresh from Ireland, or Gretchen, fresh from Germany, but of simplified electrical appliances."[95] As Edison's remarks make apparent, technological futures tied to capitalist development iterate a fantasy that machines can take over "the dull, dirty, repetitive and even reproductive labour performed by racialized, gendered and colonized workers in the past," serving to liberate a human potential "that is defined in relation to degraded and devalued others—those who were never fully human."[96]

The oft-cited etymological origins of "robot" are instructive in this regard. Ten years prior to the publication of the "The Robot's Lullaby," the word was first used to refer to automata in Karel Čapek's play, *R.U.R.*, having been adapted from the Slavic term *robota* meaning "serf" or "slave."

The robot—the automated forced-laborer—is the worker who cannot refuse. Imaginations of technological surrogacy—whether aspirational or paranoid in their declarations of freedom or replacement—are thus enmeshed with racialized and gendered hierarchies of the more and less than human.

The hierarchies of (in)humanity underpinning imaginations of technological surrogacy are also central to housewifization. For Mies, the development of the gendered division of labor is a violent process through which certain peoples are able to establish an exploitative relationship between themselves and women, and other peoples and classes: "The predatory patriarchal division of labour is based, from the outset, on a structural *separation* and *subordination* of human beings: men are separated from women. . . . the 'own' people are separated from the 'foreigners' or 'heathens.'"[97] Consequently, housewifization is one component within a larger architecture of gendered and racialized exploitation and oppression. Mies argues that the process by which women in Europe were subordinated is linked to the process by which nature was dominated by man and to "the conquest and colonization of other lands and people."[98] It is therefore no coincidence that the rise of the housewife in the Europe and North America occurs alongside the expansion of colonialism and imperialism. The naturalization and externalization of colonized peoples is the counterpart to the civilizing of European women as housewives. These two processes are not historical parallels but are causally linked within and by a patriarchal-capitalist mode of production, and are symptomatic of a two-faced European colonialism whereby some women are made "ladies" and others "savages." Thus, "While the Big White Men . . . appropriated land, natural resources and people in Africa, Asia and Central and South America" and "disrupted all social relations created by the local people, they began to build up in their fatherlands the patriarchal nuclear family."[99]

If colonization is the other side of housewifization, then the social fiction of the mystified "native" or racialized listener can be understood as the other side of the archetypical leisured white woman listener. And if the latter served to foster the consumer market for the phonograph, the former mitigated against the scrambling of racial signifiers that accompanied recorded and reproducible sound in its novel separation of sound from source: the production of "a voice without a face."[100] In disturbing the traffic between the sonic and the visual, sound reproduction technologies also disturbed the traffic between visible markers of race and sonic perception, allowing the emergence of what Alexander Weheliye elucidates as "new forms of sonic blackness."[101] For Weheliye, this perturbation is exemplified by the late nineteenth-century genre of "coon songs." Popular among middle-class audiences and with origins in the nineteenth-century minstrel show, they relied on the stability and verifiability of the performer's racial identity. Yet this was undermined by its recording insofar as the phonograph denied listening audiences, at least initially, any easy way to discern "white projections of blackness from a black image of blackness."[102]

Sound recording and reproduction's capacity to disrupt traffic between aural and visual racial markers was counterposed by rhetorical and iconographic representations affirming the ontological disjuncture between new technological objects and racialized and colonized subjects. From the 1880s through to the 1920s, written and visual depictions of racially distanced "natives" who are fooled by the sonic mimicry of the phonograph can be found in travel narratives, trade publications, and newspapers. As Lisa Gitelman explains, "The politics of these interracial encounters was characterized by the self-congratulatory aggression of Western technological achievement and colonial dominance."[103] The "magic" and "mastery" of the white man's mimetic sound technology is affirmed in the phonographic *mise-en-scène*, while serving to (re-)reproduce the colonial trade's construction of "primitive" native.[104] Elsewhere, depictions of technological ignorance made reference to the assumed amorality of racialized and colonized peoples. In the professional electrical literature, Black and Indigenous peoples were frequently cast as members of a criminal class, against which sound technologies could be used as a means of manipulation and discipline. One such story concerned the theft of several horses from Julian, California, by "a local Indian [*sic*]," who was convinced to come in and hear the "Great-Spirit" talk on a newly introduced telephone:

> The Indian took one of the cups and was thrilled with astonishment of being apparently so near the Great Keeper of the happy hunting grounds. After some little time spent in wonderment, the Indian was solemnly commanded by the Great Spirit to "give up those stolen horses!" Dropping the cup as if he had been shot, the Indian immediately confessed to having stolen the

> horses, and tremblingly promised if his life was spared, he would restore the "caballos" at once, and he did so.[105]

Whereas women listeners were also derided for their technological ignorance, they were nonetheless able to formulate appropriate relationships and uses of sound reproduction devices in the home. By contrast, such depictions of Indigenous and racialized peoples' interactions with sound technologies serve to justify the imposition of colonial power upon populations deemed naïve, amoral, subservient, and inferior. In short, they affirm the externalization of the colonized.

On the one hand, the "uncivility" of racialized and colonized peoples was thought to inhibit them from becoming properly human listeners of the early telephone and the phonograph. On the other, the labors and cultures of racialized and colonized peoples were directly involved in the early recording industry. With the emergence of the phonograph market, shellac, a resin produced by the lac beetle native to Southern and Eastern Asia and from which phonograph discs were made, followed the colonial exports of sugar, cacao, tea, and spices into US and European homes. Although it was one of the earliest transnational commodities in the seventeenth century, the shellac trade rapidly expanded at the beginning of the twentieth century due to the phonograph's rising popularity and the emerging preference for shellac discs over cylinder records. A product of capitalist extraction, the shellac records of the early twentieth-century home were reliant on the labor of men, women, and children in colonial India who harvested and processed the insect resin.[106] In addition, racialized and colonized people were called upon to produce new sonic goods for consumption: their externalization did not prevent them from entering the phonograph recording market and, from there, the white households in the metropole. The aforementioned "coon song" record market, for example, marks the popularity of a commodified, white-projected musical Blackness. Black performers were part of the phonograph and recording industry from the beginning of its mass entertainment function in the early twentieth century, becoming a substantial part of the recording industry in the 1920s, with the boom in ethnic and race record production. Indeed, it is notable that some of the respondents to The Edison Company's aforementioned 1921 survey are listeners who enjoy the way recorded sound stimulates "exotic cultures."[107]

To enter the domestic sphere, racialized and colonized peoples and their sounds were themselves in need of domestication. It was white women phonograph recordists who were often responsible for this process of domestication. In her analysis of the world music industry that emerged from the "sonic industrial revolution," Roshanak Kheshti notes the prevalence of white women comparative musicologists for the Bureau of American Ethnology at the beginning of the twentieth century, working to "salvage" disappearing Native American sounds. Social Darwinist theories at play during this time posited white women as proximate to yet hierarchically above Native and African Americans: "her race" and "her imagined predisposition to social reproduction" meant that the middle-class white woman was considered to be "uniquely capable of civilizing those African and Native Americans perceived as racially inferior."[108] Combining and affirming the role of white women as paradigmatic modern listeners and civilizers of racial and class inferiors, the white woman comparative musicologist at the turn of the century enacted the phonographic subjectification of the Native American: "when passed through the 'talking machine,' Indians were rendered legible, bought into language, organized and catalogued."[109]

Both the white woman comparative musicologist—as the first listener—and the recording phonograph—as prosthetic ear/hand—became surrogates, standing in for listeners to come. They also provided the groundwork for what became an international recording industry: Kheshti argues that the salvage recordings of female comparative musicologists were significant for the burgeoning music industry's business model, which, by the turn of the century had established offices and conducted recordings worldwide. By 1905, the Gramophone Company had made recordings in most European and Asian countries, while engineers from Victor had toured South America and the Caribbean.[110] Indeed, some of the phonograph recordings made by musicologists for archival collections made their way into homes via commercial marketing and distribution. White women recordists in the field and listeners in the domestic sphere were connected by their consumption of the aural other.[111]

Attending to the relationship between housewifization and colonization makes apparent that the phonograph's mechanization of the housewives' domestic duties amounts to outsourcing: the (partial) transition of musical and domestic duties from the housewife to the machine is predicated on a transnational infrastructure of extraction and exploitation, connecting colonized workers extracting and processing shellac to the musicians and singers captured by the phonograph company's recordists. Yet, at the same time, the phonograph's connections to colonization also enabled it to exceed the domestic sphere and bourgeois notions of reproductive respectability.

As phonograph technologies became increasingly affordable over the 1910s and 1920s, they increasingly reverberated in other sites and spaces of sociality both within and outside of the metropole. In colonial port cities, a phonograph culture emerged that, as Michael Denning observes, "did not necessarily require the purchase of a gramophone because records were heard not only, or even mainly, in the home." Instead, recorded music came to occupy "local spaces of musicking," as phonographs became public "musical instruments": recorded music was regularly heard in "bars, coffee shops, bodegas, barbershops, and in the street."[112] Indeed, the phonograph's ability to bring a wider musical culture into the home and other everyday commonplaces, combined with its facilitation of a vernacular music market, carried a destabilizing potential. While undoubtedly shaped by colonization and its musical, technological, and aesthetic priorities, Denning suggests that the boom in vernacular phonograph musics stemming from and circulating within colonial ports in the second half of the 1920s constituted a cultural revolution containing decolonial resonances. Specifically, vernacular phonograph musics contributed to a "decolonization of the ear," prefiguring political decolonization. Neither straightforward expressions of colonial culture, nor easily reconciled with a nationalist aesthetic due to their syncretic origins, vernacular musical forms such as Argentinian *tango*, Nigerian *jùjú*, Indonesian *kroncong*, and Cuban *son*, among many others, came into contact with insurgent movements:

> These musics and records shared the ambiguity of the colonial ports and railway towns in which they took shape: they were two-faced musics, objects of suspicion, intimate with both colonial and Indigenous forms and instruments. However . . . these recorded musics were quickly mobilized in anticolonial struggles, and became part of the emerging cultures of resistance. Even when these musics carried no apparent political meaning, their disruptive noise challenged not only the musical codes of empires and racial supremacy, but also the improving and uplifting ideologies of many colonial elites.[113]

The phonograph, then, was a technology of capitalist, colonial, and patriarchal reproduction, affirming the gendered separation of the two spheres and empire's hierarchies of the human. Yet, as it became increasingly affordable, the phonograph sometimes broke away from its designated social base, facilitating music cultures and practices that threatened to undermine the ideological, aesthetic, and sensory investments of colonialism.

## CONCLUSION: MUSIC AS MECHANIZATION

Often afforded a starring role in origin stories about sound recording and reproduction, the phonograph exemplifies how these ostensibly technological processes and practices are imbricated with gendered and global divisions of labor. As a middle-class consumer technology, the phonograph was expressive of and contributed to the housewife's transformation from a Victorian "angel of the home" to a "domestic engineer." Its frequent framing as a technology of leisure and entertainment, providing relief from the demands of men's waged work and women's household responsibilities, conceals its role as a technology of domestic mechanization that directly intervened in women's musical, familial, and social labors, enabling her to become a user (although not necessarily a *listener*) of recorded music. This narrative, however, requires a recognition of the housewife not as an inevitable outcome of gender difference but, like the phonograph, an industrial invention, emblematic of the separation of production and reproduction, the (partial) exclusion of women and children from the waged workforce, and the attachment of reproductive respectability to the family form. The phonograph itself reflects the growing commodification of household resources and the establishment of the home as a key site of capitalist consumption.

If the housewifization of the phonograph served to concretize its form and function, providing the technology with its initial consumer market, colonization provided the phonograph with material and musical resources for extraction and exploitation. Yet colonized subjects became listeners, consumers and users too, as the phonograph and its records became more affordable and manufacturing expanded outwards from the colonial core. Here, the phonograph entered not only the home but also communal spaces of sociality and life making. Indeed, sound reproduction's entry into the domestic sphere was never exclusive: the technology led parallel and sometimes contradictory lives, appearing as a cheap form of urban, working-class entertainment, as well as a musical instrument for the refined home; a product of colonialism and a transmitter of anticolonial resonances.

As will become apparent over the following chapters, the construction of the phonograph as a "mothering machine"—a means of automating, assisting, and enhancing the housewife's domestic duties—preempts the design and circulation of many of the sound technologies called upon to ameliorate the crisis of reproduction as it appears a century later. Of significance too is the functional presentation of recorded music and the accompanying distinction between music's listener and user. Both are reiterated with later reproductive sound technologies. Thus, although the material-ideological constitution of the family, mother and wife has shifted significantly since the early twentieth century, the notion of the sonic surrogate persists.

# 2 "Hear for You"

## SMART SPEAKERS, EXTRACTIVE LISTENING, AND TECHNO-AUSTERITY

In a 2019 advert titled "Sharing is caring," a dutiful grandson brings his disinterested grandfather a new generation Echo Show: an Amazon smart speaker featuring a touchscreen display.[1] Launched in 2017, the Echo Show, like many of the other voice-controlled devices in Amazon's product range, functions simultaneously as a home entertainment, communication, and home automation system. It enables users to listen to music, radio, and audiobooks; make phone calls; set timers, alarms, and reminders; schedule daily routines; ask for wellness tips; purchase items from Amazon's e-commerce platform; and operate compatible devices such as lights, central heating, and door locks. Accordingly, the grandson in "Sharing is caring" enthusiastically tells his grandpa—and the advert's audience—that he can use the Echo Show to "check the weather, or listen to some music." Grandpa, however, remains more interested in the archaic medium of the print newspaper. In the next scene, the grandson repeatedly sings the name "Valerie" to himself as he washes up, before instructing the device's digital assistant Alexa: "At 6 p.m., remind grandpa your favourite dessert is in the fridge."

By the grandson's next visit, a relationship has been forged between the smart speaker's digital assistant and its older user. As his grandson is seen walking down the path to his home, grandpa asks Alexa to "play that song that goes 'Valerie.'" For listeners familiar with the different renditions, the grandson's melismatic delivery heard prior makes clear that it is Amy Winehouse's neo-soul vocals from Mark Ronson's cover version that he has in mind. Alexa appears to know intuitively which version of this song to play and immediately obeys Grandpa's command. Winehouse's distinctive voice fills the room as the grandson walks through the door. While he might need a bit more help around the home, grandpa, it would seem, is still young at heart.

Smart speakers are an emerging consumer technology market: in 2025, it was reported that 35 percent of the US population aged sixteen-plus and 45 percent of the UK population aged sixteen-plus own at least one smart speaker.[2] Since the launch of its first Echo speaker in 2014, Amazon has come to be the leading vendor of these devices.[3] The Echo range is one of the multinational company's multiple forays into consumer hardware: other products include the Kindle e-reader, Kindle Fire tablet, Amazon Fire TV (a line of video streaming players and consoles, some of which have a smart speaker integrated) and the short-lived—and notoriously unsuccessful—Amazon Fire phone.

Referring to the Echo as a speaker, however, risks concealing aspects of its functionality. Where speakers are typically associated with sound's transduction, production, and playback, the *smart* speaker aims to facilitate a two-way exchange of auditory information. Since the Echo's primary purpose is as a mediator of Amazon's digital assistant, Alexa—a cloud-based, AI voice service—it contains a series of microphones, enabling the device to record as well as broadcast sound. As well as the Echo range, Alexa is a feature of other Amazon-branded devices such as its Fire TVs and tablets. It is also increasingly integrated as a third-party feature into other manufacturers' products: for example, Alexa features on certain models of Bose headphones, Sonos soundbars, Fitbits, and Guess watches; and is downloadable as an app on Android, Microsoft, Apple, and Google platforms. At the beginning of 2019, Amazon claimed that it had sold 100 million devices with Alexa on board.[4] This expansion of products featuring smart speaker capabilities has also resulted in Amazon's devices moving beyond the domestic domain, with which they are primarily associated. There is now, for example, Echo Auto, which enables users to add

Alexa to their car and "can hear you over music, air-conditioning and road noise"; while the integration of Alexa into Fitbits and watches allows the assistant to accompany users "on the go." Alexa for Business, meanwhile, promises to enable employers and organizations to "get more done" and "lets your employees be more productive throughout their day at work, at home or on the go."[5] These different devices and services speak to an aspiration to make Alexa ubiquitous.

Commentators have frequently remarked upon the gendering of smart speakers and their digital assistants as feminized technologies. This feminization is understood as a product of both what the device does and how the device sounds. Focusing primarily on the digital assistant, Yolande Strengers and Jenny Kennedy, for example, have suggested these technologies are presented as "the smart wife." Friendly and constantly available, they are imagined to undertake "wifework"—that is, the managerial and affective activities associated with the wives and mothers in the home. For Strengers and Kennedy, the prototype of these devices, which originate in male-dominated design environments, is the archetypal 1950s housewife: white, middle-class, heteronormative, and simultaneously responsible, organized, and nurturing.[6] To this end, these devices might be considered a progeny of the phonograph and its housewifization, outlined in chapter 1.

Strengers and Kennedy's assessment of the device's association with "wifework" is seemingly affirmed by another 2019 advert, which depicts an Echo device helping a new father taking care of his young daughter. The speaker provides the new dad a series of reminders from "Laura": the teething ring is in the freezer, there's a play date at 3 p.m., and finally, "Laura loves you."[7] The advert is by no means unusual: scenes of middle-class domesticity, attached to the reproductive logics of the nuclear family, are recurrent tropes in promotional materials for Echo devices. Yet the smart speaker and its digital assistant are also reflective of a blurring of the spatial and temporal boundaries of home and work, labor and leisure. The description of voice-operated software as "assistants," for example, makes apparent their connection to the administrative tasks of personal assistants: as the advert makes clear, they can help users to manage their calendars and provide helpful reminders, alerts, and source information, as well as act as a maternal proxy. The smart speaker thus combines professional and intimate formations of gender.

Although advertising often positions them amid familial relations, the conflation of smart speakers and their digital assistants with the figure of the housewife risks glossing their relationship to other histories of domestic labor. Thao Phan has offered an important corrective to Strengers and Kennedy's account, arguing that the focus on the figure of the housewife in analyses of these technologies can obscure hierarchies of labor and power that occur between women within the domestic sphere. For Phan, digital assistants reference an idealized figuration of domestic service, predicated on the historic and ongoing racialized exploitation of domestic laborers by wealthier households. It is not the 1950s housewife but the servant that is the device's prototype, with users being able to issue commands and have their needs met.[8] The framing of the Echo Show in the "Sharing is caring" advert, where it is figured as a personal care assistant for an older user, lends further weight to Phan's corrective. Broadcast in the US, Canada, UK and some other European countries (including Germany and France), "Sharing is caring" is reflective of wider trends in the organization of care in advanced economies. While responsibility for the care of the disabled and the elderly is often shouldered by individual families (as alluded to by the presence of the grandson in the advert), as a form of waged work it is largely performed by working-class women, migrant women, and women of color: like the domestic servant, it is a role predicated on racial and class-based divisions of labor within the home.[9]

The feminization of the smart speaker is reinforced by its "voice." Amazon's Alexa, for which the Echo speaker is a mediator, is one manifestation of a historical chorus of speech synthesis technologies that are programmed as "female."[10] The feminized voice of the digital assistant communicates affective openness, being heard as warm, friendly, and attentive. This voice is also racialized. As made apparent in chapter 1, sound reproduction technologies both reproduce and reconfigure racial formations. Where the work undertaken by digital assistants is indebted to racial histories and fantasies of domestic service, this is partly concealed by an oral aesthetics of whiteness. As Phan makes clear, Alexa's default setting is to speak Standard English using a "neutral" US accent, in a way that appears mannered, educated, intimate and yet professional. While the servant/master relation has been "fundamentally predicated on racialized human exploitation," this is effectively erased by the figure of the white-speaking, middle-class AI. Phan suggests that the (de)racialization of the smart speaker's digital assistant in accordance with the norms of whiteness enables discomfort associated with the racialized histories of servitude to be mitigated, while also likely reflecting how whiteness has been central to understandings of "universal" subjecthood.[11] What the smart speaker says and what it does are thus predicated on at least two different processes of feminization that are distinct in terms of race and class. Whereas the digital assistant's archetypal voice and proximity to administrative work references white, educated femininity, their positioning in relation domestic service and personal care also feminizes these technologies through reference to the oft-abject work of migrant women, women of color, and working-class women.

There is, however, an additional yet rarely acknowledged dimension to the smart speaker and its digital assistant that contributes to their feminization. In an Amazon review of the second-generation Echo Dot from February 2018 titled "Divorced Siri," the device's digital assistant is described as the "woman" superior to Apple's competitor. In keeping with the blurred distinction between the professional and the intimate, the reviewer switches between referring to Siri and Alexa as employees (who can be fired) and romantic partners (who require "uncoupling"). The reviewer writes:

> So, I fired Siri today. I'm not sure she understood, for she has never understood me. So, we did an "uncoupling." . . . I tried explaining that Alexa *really listens to me*. . . . Again, Siri said she couldn't understand me. As Alexa and I grew in our new relationship, she now takes care of most of my needs. . . . I sit on my couch, Alexa orders me take out, she checks to make sure garage door is closed, . . . she "drops in" the kitchen Echo to help the kids check math homework. *If anyone wants something at home that actually listens to you and does what you ask, Alexa is your girl.*[12]

It is the reference to Alexa or, in actuality, the Echo smart speaker as *listening* that I want to draw attention to: it is the smart speaker's capacity to "listen" that is central to its gendered operability, yet this has often been overshadowed by a focus on the device's vocal attributes. Responding to the absence of listening in literature about smart speakers, digital assistants, and gender, as well as to an absence of labor in literature about listening, in what follows I propose that listening offers a productive conceptual framework through which to analyse the smart speaker, enabling connections to be drawn between various aspects of its production, design, and circulation. Taking Amazon's Echo and Alexa as exemplary, I describe the smart speaker as listening in the sense that, as speech recognition technologies, they capture, process, and aim to appropriately respond to auditory information. They are also listening devices in the sense that they are modeled on the auditory relations associated with—and constitutive of—feminized work. Certain practices and ideals of listening are attached to the various archetypes of feminized work commonly associated with the smart speaker: the personal assistant or secretary, the servant and the carer, as well as the housewife. Put simply, it is not just what the smart speaker does but *how it listens* that connects the smart speaker to feminized work. Yet smart speakers and their digital assistants are not reducible to the promises of automation, gendered fantasies, and familial imaginaries that are attached to them. An attention to listening can help reveal the limits of conceptualizing these technologies in relation to feminized work, while also making apparent the coherence between the smart speaker's gendered presentation and underlying economic logics.

On the occasions when the capacity of the smart speaker to listen has been mentioned by cultural commentators, it is often as part of anxious narratives about privacy and surveillance. Questions therefore arise as to how users are incentivized to engage with a device that has—via sound—a great deal of access to their intimate lives. In response, I identify three intersecting factors that have driven adoption of the smart speaker: first, the cheapness of these technologies; second, the presentation of surveillance as service, aided by both feminization and the naturalization of aurality; and third, austerity and its associated crisis of social care. Highlighting the smart speakers' use by English Local Authority care providers, I suggest that the device has become a stop gap for the problems created by, on the one hand, increased demand and, on the other, chronic underinvestment, underfunding, and workforce shortages in social care. The language of cuts, crisis, and shortages and through which austerity is ordinarily articulated, however, only tells part of the story: Amazon's insertion into publicly funded care services makes apparent that, for some, austerity has its benefits.

## CAREFUL LISTENING

In both feminist theory and sound studies, listening has been valorized as a means of connecting to the world and one another. Its capacity to facilitate different modes of ethical understanding and responsibility and its proximity to dialogue, relationality, and openness has meant that listening is recognized as tool of revelation. Salomé Voegelin, for example, argues that while listening will not automatically provide us with a better world, it can nonetheless be politically and epistemologically useful in revealing the "world in its invisibility: its unseen movements beneath its visual organization." A philosophy of sound, meanwhile, "*takes care* of the audible as the possible, the 'what could be,' or indeed the 'what there is' *if we would only listen.*"[13] These revelations, however, are thought to come from a position of vulnerability. Brandon LaBelle, for example, describes listening as "a condition of weak-strength" inasmuch as the "dialogical moment" requires "always listening beyond myself, moving my own views in consideration of another's, giving my attention to opinions different from my own while finding ways to resist and counter their power if need be."[14] As both these descriptions suggest, an open, vulnerable, and careful listening that makes apparent our relationality and responsibilities is deemed ethically, politically, and ecologically valuable.

Seeking to counter Western political philosophy's emphasis on the voice as an expression of political personhood, feminist scholars have highlighted the centrality of listening to social justice, consciousness raising, and emancipatory politics.[15] Writing on listening's relationship with feminist care ethics, for example, Sophie Bourgault emphasizes that the refusal to listen can itself be a manifestation of oppression. Conversely, being listened to can be "profoundly empowering." Noting the value of listening in educational and healthcare settings, Bourgault identifies "attentive listening" as central to the pursuit of recognition and social transformation. Such listening is under threat: with the establishment "neoliberal learning environments," Bourgault suggests that there is a "disquieting trend in terms of the cultivation of social skills and in particular, of dialogical listening skills" insofar as "bodily presence adds an additional texture and meaning to our listening, as does shared attention."[16]

Listening, then, is considered crucial for generating relations of care and for understanding social, environmental, and affective worlds. However, while many of its advocates are careful to situate it as a cultural and historical practice, there remains a tendency to romanticize listening and "ontologize" the ethical hopes attached to it. Meanwhile, listening's implication within gendered, racial, and international divisions of labor and, with it, a more ambivalent auditory politics of care have been comparatively sidelined. The repeated assertion that listening brings forth the new or allows us to hear otherwise is counterposed by the ways that listening—particularly "good," empathetic, and attentive listening—is both idealized and exploited in relation to various forms of feminized work. This is true of professional roles such as counselors and educators; service and customer-facing roles, such as shop assistants, waiting staff, call-center operators, and flight attendants; and caring roles such as nursing and home care assistants.

To describe these roles as feminized is not simply to comment on the gendered makeup of their workforces but also their proximity to qualities, skills, and attributes historically associated with normative constructs of femininity, such as mediation, flexibility, adaptability, and the management of affect. While it may be undertaken by workers of different genders, the apparent compatibility between feminized work and feminized worker can serve to naturalize certain skills they bring to and draw upon in the role, justifying low levels of remuneration. For example, certain types of health and social care work are low paid because the skills involved are imputed to an ascribed identity trait: women—particularly women of color—are envisaged as inherently caring or nurturing, thus the service they provide is considered an extension of their intrinsic skill rather than the cultivation of professional competency.[17] However, not all workers are gendered equally. Some feminized sectors have been historically and are increasingly reliant on the cheap and flexible labor provided by migrant workforces—particularly customer service, domestic service, and health and social care—with workers being exported and imported as part of transnational networks of capital and labor.[18] Feminized work, then, is a heterogeneous category containing significant racial, national, and cultural differentiations, while the assignment of feminine qualities and attributes that are deemed to make workers more and less suited to particular aspects of feminized work often confirm hierarchies of race and nationality.

Listening has itself been naturalized as a specifically feminine attribute, which is drawn upon in feminized work and helps to define it as such. However, the types of listening that are undertaken and the extent to which listening is acknowledged as central to the accomplishment of different roles varies. For some managerial and professional roles, listening is explicitly named as a valuable "soft skill," while for educators, therapists, and counselors it is a fundamental component of their employment. In other types of work, however, being a "good listener" or "easy to talk to" may make one a more desirable employee or more effective at a role, even though it is not formally part of the position. For home care assistants, for example, being able to lend a listening ear is often an unacknowledged but central component of their work. In their study of Canadian home care, Jane Aronson and Sheila Neysmith note that care agencies and policymakers tend to focus on the practical aspects of home care assistants' roles, which consist of assisting service users with personal care (bathing, washing, dressing, eating, and monitoring health conditions) and household chores (cooking, cleaning, shopping, laundry). However, in contrast to these formal accounts orientated towards practical activities, home care assistants' own accounts of their work draw attention to its affective and communicative dimensions, and, by extension, the social bond between caregiver and receiver. This includes talking with and listening to clients. The disjuncture between agency and employee understandings of the role can lead to competing demands: the desires of home care assistants to listen to and engage in conversation with their clients, for example, exists in opposition to a trend towards depersonalization, in the interest of making care work more efficient.[19] Listening, then, serves to aid the home care assistant in the work, while also existing in tension with the push to reduce labor time.

Other domestic workers have been expected to undertake listening work in addition to or alongside their formal role within the household. In her 1985 ethnographic study of the relationship between women within the home, Judith Rollins details instances in which women employers would approach domestic employees as confidantes: "Some domestics heard details of their employers' extramarital affairs; many heard about strains in employers' marriages."[20] The physical closeness and social distance between employer and worker—exacerbated by racial difference—meant that employers felt free to tell domestic employees secrets they would not share with friends or family. In the words of Jane Louis, a former domestic worker and one of Rollins's interviewees, "Most employers like to talk to the people who work for them as you're not in their circle, you're not going to tell anyone who is important to them. . . . They talk to you anyplace. A white person will go up to a black stranger and tell them very private things—because they know its not going to go."[21]

The confidential listening undertaken by domestic workers, however, is one of a range of listening modes that they may be expected to adopt. Writing on the history of domestic service in the USA, David Katzman argues that in the early twentieth century, "one peculiar and most degrading aspect of domestic service was the requisite of invisibility." Consequently, the ideal servant would be silent but also selective in their listening: they would be responsive to demands but fail to register gossip, household chatter, and conflicts; they would be attentive to the needs of their mistress and master but overlook their faults; and they would be "sensitive to the moods and whims of those around them but undemanding of family warmth, love or security."[22]

While the working conditions of domestic workers undoubtedly changed between the early and late twentieth century, the desire for invisibility endures. Reflecting on her own employment as a domestic worker, Rollins notes how to one employer, "I became invisible; their conversation was as private with me, the black servant, in the room as it would have been with no one in the room. . . . These gestures of ignoring my presence were not, I think, intended as insults; they were expressions of the employers' ability to annihilate the humanness and even, at times, the very existence of me, a servant and a black woman."[23] Listening selectively—to commands but not conversation—is a constitutive component of domestic invisibility, often predicated on racialized dehumanization. Where the domestic servant may sometimes become confidante or interlocutor, their listening is to be activated by an employer.

Listening serves to connect the various feminized roles that are traceable in relation to the digital assistant: it is a requirement of the personal assistant or secretary, the servant and the carer, as well as the housewife. While there remain important racial and class-based distinctions between these figures, they cohere around an expectation for attentive, empathetic, careful listening. This idealized and feminized listening informs the design and deployment of smart speakers such as the Echo: they are there to listen to user's administrative requests, logistical commands, and potentially prying questions, remaining unobtrusive but always alert. However, in a manner akin to the "invisible" domestic worker, the smart speaker's listening should be selective, discreet, and activated at the user's request. In other words, the operations of the smart speaker are modeled upon the social *and* sensory relations of feminized work.

## THE EMOTIONLESS COMPANION

In March 2017, Amazon Technologies Inc. filed a United States Patent for "voice-based determination of physical and emotional characteristics of users." The patent states, "Certain content consumption devices may be configured to receive voice-based commands, or may otherwise be configured to recognize speech. Voice input from users to such devices may reflect a physical or emotional characteristic of the user. Accordingly, determining a physical or emotional characteristic of a user using a voice input may be desired."[24] Accompanying the patent is an image depicting a friendly conversation between woman and machine, with the latter once again occupying the role of caregiver (figure 6). The user, who is depicted as coughing and sniffing, tells Alexa that she is hungry. The offer of a recipe for chicken soup is declined. However, when Alexa, registering the user's ill-health, proposes to order some cough drops with a one-hour delivery time, the cylindrical speaker is told: "That would be awesome, thanks for asking!" The patent describes numerous other scenarios in which the devices' capacity to attune to the user's emotional and physical state would be advantageous for both prospective advertisers and users: a bored disposition may lead Alexa to ask, "Are you in the mood for a movie?"; the device may "match" the voice of a "sleepy" user with a targeted audio ad for a singer's new album. In processing the user's vocal data—"listening out" for "pitch, pulse, jittering, voicing and/or harmonicity"—Amazon's voice-activated technologies may be able to recognize "joy, anger, sorrow, sadness, fear, disgust, boredom, stress, and other emotional states."[25] The user's voice not only offers instruction but is a mediator of affect, from which Alexa's microphonic "ear" can glean potentially beneficial insights.

These patented developments ostensibly mark an attempt to improve the listening capacity of Amazon's smart speakers, as well as coinciding with multiple ventures into the healthcare sector. These include

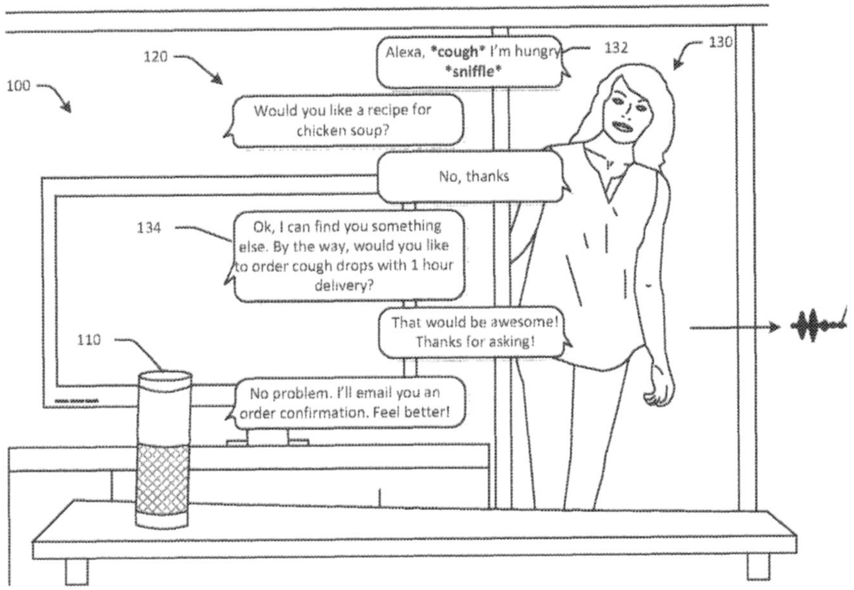


*Figure 6.* Illustration from Patent US010096319B1, Voice-based determination of physical and emotional characteristics of users, 2018.

partnerships with pharmacy chains Coco Kara Fine Inc. and Matsumotokiyoshi Holdings to create a delivery service for Japanese users; and the establishment of Amazon Care, a pilot app launched in 2017 that enables users in Washington State to access virtual and in-person healthcare services, following the acquisition of e-health startup Health Navigator. However, while the predictive capabilities of "sentiment analysis" have been promoted for many years by digital data companies, it would be erroneous to suggest that smart speakers and their digital assistants are currently capable of "reading" the health status or emotions of a user.[26] Indeed, the model of interaction detailed in the patent is based upon a device's programming registering and responding to overt sonic cues (e.g., a cough), rather than the establishment of an empathetic AI. Nonetheless, the aspiration for these devices to register what the user is saying and *how* they say it serves to direct attention to the listening smart speaker's association with affect and the production, presentation, and interpretation of feeling.

The generation, transmission, recognition and modulation of affect is considered integral to both feminized work and digital media technologies. Analyses of affect in both these domains are frequently indebted to Arlie Hochschild's sociological study *The Managed Heart*.[27] First published in 1983, Hochschild theorizes the commercialization of emotional labor and the accompanying "transmutation" of private feelings into economic circulation. Centering her analysis on the service work of flight attendants and the bill collector, Hochschild examines the "feeling rules" that guide differently gendered and classed workers in managing their moods and emotions and, with them, the moods and emotions of others.[28] This kind of labor, Hochschild argues, "calls for a coordination of mind and feeling, and it sometimes draws on a source of self that we honour as deep and integral to our individuality."[29]

For Hochschild, commodified emotional labor is predicated on alienation. Drawing comparisons between the physical labor of the industrial factory worker and the emotional worker of the service economy, Hochschild notes that for both, "the worker can become estranged or alienated from an aspect of self—either the body or the margins of the soul—that is *used* to do the work. . . . If we can become alienated from goods in a goods-producing society, we can become alienated from service in a service producing society."[30] While Hochschild is clear that all emotional labor, whether alienated or unalienated, is socially constituted and mediated, the management of feeling risks cutting those who undertake it off from the "unmanaged heart."[31] There are times, however, that workers resist the alienation demanded by commodified emotional labor. Reflecting upon descriptions by airline attendants of what happens when the depth of feeling required to effectively perform emotional labor becomes untenable, workers are described as going "into robot": they make clear that they are withholding any genuine feeling and become perceptibly superficial in their emotional display.[32] In adopting a robotic mode, airline assistants refuse to put their "full" selves into their emotional labor: they withhold access to true emotion.

Emulating the "feeing rules" associated with care, domestic, and administrative work, the smart speaker and its digital assistant incorporate affective cues of politeness and hospitality. Whereas Hochschild's analysis primarily focuses on the visual displays of feeling—the smile of the flight attendant is a recurring theme—the disembodied digital assistant displays affect sonically, via the AI's voice. Alexa remains warm and courteous even when faced with impossible questions, offering apologies when commands and requests are not understood. However, Alexa does not only represent affective cues: smart speakers have the potential to modulate feeling by forging affective relations between user, assistant, and the digital content they provide. In a small study of older users of digital assistants and smart speakers, Margot Hanley and Shiri Azenkot note that an emerging theme concerns their capacity to alleviate feelings of loneliness. One interviewee—a woman in her 80s—states, "It's just the idea that you're not alone, . . . you just got somebody there." Another female interviewee in her 80s suggests, "It also becomes like a friend. . . . You don't feel so . . . living in silence," while her husband, states, "I don't know whether it's a friend or not a friend, but you're not alone, even if, you know—because you can just reach out verbally and get music."[33] These reflections on users' own affective bonds with smart speakers and their digital assistants resonate with Amy Schiller's and John MacMahon's suggestion that as a kind of "emotional companion," Alexa builds new feeling rules for the marketized home. In so doing, Schiller and MacManon suggest that household AI such as Alexa disrupt the dualisms of emotion/human and robot/automation implicated in Hochschild's analysis.[34]

For other critics, too, Hochschild's binary of human and nonhuman, and the associated dualisms distinguishing alienated and unalienated affective states have been a source of contention. Kathi Weeks argues that although Hochschild is aware of the conceptual limitations of alienation as it was developed in humanist Marxism (namely, a tethering of critique to a nostalgic ideal of preindustrial work) and is careful to avoid advocating for an authentic emotionality, her analysis remains reliant on a pregiven self, one that is estranged through emotional labor. When Hochschild highlights the ways in which the subjectivity of the emotional laborer is cultivated, a distinction remains between who we are in work and who we really are. Consequently, Weeks argues that there is a tension at the heart of Hochschild's analysis: "she insists on the social construction and malleability of the emotions while also positing them as fundamental to the self such that alienation is a problem."[35]

Building on Weeks's critique, Shiloh Whitney argues that Hochschild's comparison of emotional labor and industrial labor obscures key differences between the production of goods and the production, circulation, and management of affect. For Whitney, affect cannot be sufficiently understood in terms of the distinction between authenticity and inauthenticity, true and fake feelings. To assume such a distinction is to obscure the ways in which outward displays of affect carry to others and the environment and, furthermore, affect the feelings of workers themselves. Affects do not simply move from worker to consumer; they are not used up or exchanged but transmitted and amplified across, between and within bodies. Indeed, "affective labour is impressively effective at reconstituting the boundary between feelings that belong to me and feelings that do not."[36] Purposely sidestepping the distinction between productive and reproductive labor, Whitney suggests that affective labor can be more effectively conceptualized as *byproductive* insofar as the management of feeling involves not only the production and transmission of affect but also the metabolization of waste or excess affect. Such is the case when domestic workers, for example, have to "swallow their pride," choke back anger, or absorb anxiety.[37] These affective processes call into question the separations assumed by Hochschild's description of emotional labor as alienated.

Whitney's description of affective labor as a social relation that involves not only the production, transmission, and modulation of feeling but also the metabolization of waste or excess affects is echoed in the oft-reported tendency for users to insult and verbally abuse smart speakers and their digital assistants, approaching these devices as a *container* for frustrations and bad feeling. In a 2018 Wired article entitled "The Terrible Joy of Yelling at Alexa," Emily Dreyfuss describes how she "fell in love with Alexa" after giving birth to her son. The device "was like an extension of me, doing the things I wanted to but couldn't. Sometimes my husband and I joked that she was the other wife in the family." As the demands of life, work, and parenthood became more challenging, Dreyfuss suggests that her relationship with the device changed: "The house was full of tension, but Alexa didn't seem to notice. She answered every question in the same perky tone." The robot became "a convenient target," and Dreyfuss and her husband could take out their frustrations by yelling at, berating, and insulting the device.[38] The extent to which such practices are actually common among users or are merely a construction of the North American commentariat is unclear; however, the notion that Alexa and the smart speaker can allow users to unburden affect in the home has come to be a recurrent media trope. In *The New Yorker*, Sarah Larson describes listening to music with Alexa: Larson suggests that "yelling, as opposed to poking around on a screen, can help put you in touch with your id. And it adds to the pleasure."[39] In *The Washington Post*, meanwhile, an interviewee describes how, during the COVID-19 pandemic, Alexa became a "virtual punching bag."[40] The piece identifies users' propensity to "vent" at smart speakers, particularly when they fail to meet expectations of smooth and effortless functionality—the frequent occasions when, for example, Alexa doesn't understand, or when a user command results in no response.

In these accounts, emphasis is placed on Echo/Alexa's nonhuman status: it is precisely because the smart speaker is an *un*affected, technological object and is thus able to listen without judgment or feeling, that makes it the ideal recipient of unwanted emotions and moods. As Dreyfuss puts it: "I bought this goddamned robot to serve my whims, because it has no heart and no brains and it has no parents and it doesn't eat and it doesn't judge me or care either way."[41] While the smart speaker's emulation of human patterns of communication, behavior, and activity enable it to appear as a household interlocutor, the ability of the smart speaker to engage in these affective processes is predicated on the partial maintenance of a human-machine distinction. This dialectic of humanization and dehumanization enables the smart speaker to become a listening but *emotionless* companion.

## EXTRACTIVE LISTENING

In her highly influential feminist text, "A Cyborg Manifesto," Donna Haraway contrasts the ubiquity and apparent immateriality of digital technologies with the opacity of human laborers who become part of their circuitry:

> Modern machines are quintessentially microelectronic devices: they are everywhere and they are invisible. . . . Our best machines are made of sunshine; they are all light and clean because they are nothing but signals, electromagnetic waves, a section of a spectrum, and these machines are eminently portable, mobile—a matter of immense human pain in Detroit and Singapore.[42]

Revisitations of Haraway's speculative and ironic account of boundary transgressions have typically focused on the transformation of the human-machine distinction from the perspective of the former and its implications for question of identity.[43] However, in bringing to the fore the "human pain" implicated in the "clean and light" microelectronic devices, Haraway's manifesto also undermines the automation-labor distinction. It is not just that the machine destabilizes the boundaries of what was once thought of as the human; the human also destabilizes the boundaries of the machine. The seemingly immaterial technologies of signal and waves that appear to run by themselves are tethered to the factory workers of Detroit and Singapore or, more commonly today, of cities across China, India and Vietnam.

Although smart speakers undoubtedly reference different types of feminized work in their design and circulation as consumer technologies, there remain important distinctions between the archetypal feminized worker and its machinic reimagination, including the types of listening they undertake. The recurrent comparison of the digital assistant to housewives and secretaries (and, less frequently, servants and carers) risks omitting the "human pain" that exists amid the smart speaker's soundwaves and signals, obscuring the device's status as both a commodity produced by labor and a mediating technology through which labor occurs. This bears out in many feminist accounts of the smart speaker. When labor is mentioned, it is as that which the smart speaker is modeled on or that which is automated by the device: there is rarely a consideration of the multiple types of human labor implicated in the device.[44] This collective omission, combined with Haraway's reminder of the human in the machine, encourages us to recognize that it is not simply the smart speaker we call Echo nor the voice we call Alexa that listens: rather, this listening is constituted by various labor processes that take place across the world in mines, factories, offices, and indeed the home.

If the phonograph makes apparent the imbrication of "colonization" and "housewifization," as was outlined in chapter 1, the enduring effects of this history are traceable in contemporary reproductive sound technologies such as the Echo, which combine feminized ideologies of care and domestic servitude with neocolonial relations of labor, marketization, and extraction. As Kate Crawford and Vladan Joler describe it, "each small moment of convenience—be it answering a question, turning on a light, or playing a song—requires a vast planetary network, fuelled by the extraction of non-renewable minerals, labour, and data."[45] From this perspective, automation once again appears as outsourcing: the feminized work ostensibly undertaken by the smart speaker is enabled and facilitated by this "planetary network." Predicated on both the privatization and exploitation of the Earth's natural resources and the exploitation of human labor, this network includes those involved in the manufacture and assembly of these devices: the low-paid (and potentially forced) miners, smelters, and refiners who retrieve and treat minerals in dangerous and environmentally deleterious conditions and component manufacturers and assemblers, who risk exposure to hazardous chemicals in repetitive and stressful work environments.[46]

At the time of writing, Amazon's smart speakers are assembled by Foxconn in Hengyang, China, the world leader in high-end electronics manufacturing. Founded in Taiwan in 1974, the company has around one million employees, the majority of whom are Chinese rural migrant workers. While Apple is Foxconn's largest customer, other clients include Google, BlackBerry, Dell, Microsoft, IBM, Sony, Huawei, Xiaomi, Nintendo, Panasonic, and Toshiba. The company has been accused of upholding various illegal work practices and rights violations, while Foxconn workers have resisted the company's deleterious employment conditions with strikes, protests, and work slowdowns.[47] The smart speaker's blurring of the social, spatial, and temporal boundaries that historically distinguished home and work is mirrored in the conditions of its assembly: Foxconn workers sleep on site in factory dormitories and eat in factory canteens.[48]

Once manufactured, the functionality of the Echo is similarly reliant on various labor chains and networks. This includes the labor undertaken by an international workforce of contractors and Amazon employees, who "listen in" on recordings of user's commands, interactions, and daily activities to help the machine learn. Thousands of reviewers transcribe and annotate recordings that are then fed back into the software to help eliminate gaps in Alexa's "understanding" of human speech and help the AI model to respond better to commands. According to employees, the workday lasts around nine hours, during which a thousand audio clips may be processed. These recordings are said to be taken from when the device is activated through the wake word (usually "Alexa"), though reviewers have suggested that recordings may occur when users believe the device is inactive.[49] While many of the recordings are mundane, reviewers have reported "overhearing" encounters that are distressing and upsetting. As a result, aspects of the reviewer's role cohere with Shiloh Whitney's notion of byproductive labor. Although Amazon claims it has procedures in place for workers who hear something distressing, two employees based in Romania claim that, after requesting guidance in such cases, they were told it was not Amazon's job to interfere. Instead, when reviewers experience distressing recordings, they may share the experience in an internal chatroom as a way of relieving stress.[50] As a result, the reviewer's processing of recordings so as to help Alexa "learn" is accompanied by the processing and metabolization of negative or excess affects that may arise from them.

The work of Amazon's reviewers makes clear that the transformation of the "speaker" into the "smart speaker," which appears to listen and respond to users, is predicated on the device's ability to record as well as broadcast sound. As I discuss further later, the Echo's incorporation of recording technology has been controversial. However, the importance of recording to the device and its functionality, in the sense both that it enables the smart speaker to "listen" and enables the creation of training datasets for Alexa, suggests that user interactions are themselves partly constitutive of the smart speaker's listening.[51] As Crawford and Joler argue, the key difference between AI systems and other consumer technologies is that the former relies on "the ingestion, analysis and optimization of vast amounts of human generated images, texts and videos."[52] The Echo user's voice and the commands it mediates are themselves part of this process of ingestion that serves to power AI.

A recurring theme in analyses of digital culture concerns the ways that users participate in "free labor" or "consumer labor." This free labor has at least three overlapping facets. First, free digital labor is used to refer to the generation of content that populates blogs, social media channels, music streaming services, and news sites; the development and maintenance of mailing lists and websites and the modding (i.e., modifying by third-party or user) of software and games. These technical, affective, and cultural activities are the focus of Tiziana Terranova's influential theorization of free labor, which she describes as voluntarily given and unwaged, enjoyed but also exploited: it is a manifestation of the increasingly blurred territory between production and consumption, work and cultural expression.[53] Second, free digital labor may pertain to the provision of attention.[54] The free labor of attention may be "actualized"—manifest in Facebook likes, YouTube comments, or streaming time bars. Or it may be manifest as a potential or capacity that, as the Amazon patent makes clear, helps to attract advertisers to platforms. Third—and most pertinent for understanding smart speakers—free labor within the digital economy can pertain to the generation of data. User data may drive the functionality and development of different platforms and technology; it is monetized through exchanges with third parties such as advertisers, ad servers, insurers, credit agencies, or data aggregators; or it may constitute an asset, insofar as user data contributes to a platform's market valuation, helping to attract capital investment or acquisition.[55] Of course, users are not compensated by digital platforms for the data that their interactions and engagements produce. Data is perceived as a natural resource that can be "mined" for benefits. And, as with other useful, ostensibly natural resources, data's status as such conceals the social labors upon which it is reliant.

The accrual of data from user activity or consumer interactions is central to Amazon's operations. As James Thomson, a former Amazon executive puts it: "they happen to sell products, but they are a data company. . . . each opportunity to interact with a customer is another opportunity to collect data."[56] The status of Amazon as a data company is confirmed by the centrality of its subsidiary Amazon Web Services to its operations. Established in 2006, Amazon Web Services is a cloud computing platform. It was first a means of facilitating the companies increasingly complex logistics; however it is now an on-demand rental service, offering servers, computing power, storage, operating systems, data processing, software, and applications to third-party firms. The provision of digital infrastructure enables Amazon to continually extract and expand its access to data, while also extracting rent from contemporary businesses.[57] Amazon Web Services is currently the most profitable segment of the company: in the first quarter of 2020, it accounted for 77 percent of Amazon's total operating profit; while in 2019, it generated $35 billion in revenue for Amazon.[58]

It is in this context that Echo and Alexa need to be understood. However, while Amazon's accumulation of data is no secret, the specifics of how Amazon uses and monetizes it are less apparent. Amazon Europe's privacy notice states, "we are not in the business of selling our customer's personal information to others," though the privacy notice also states that Amazon "shares" customer's personal information with "third-parties involved in transactions" and that "customer information" may be transferred as a business asset with the acquisition or selling of services.[59] But personal information is not the only form of user-generated data. In Amazon's section on "interest-based ads," the company notifies users as follows: Amazon "uses information such as your interactions with Amazon sites, content or services," and some third parties may provide "non-personally identifiable information about you (such as demographic information or sites where you have been shown ads) from offline and online sources that we may use to provide you more relevant and useful advertising."[60]

Written to assure consumers, these documents gesture to some of Amazon's (known) uses of customer data. The accrual and analysis of data is central to Amazon's internal operations, including for predictive purposes such as forecasting market behavior with a view to refining supply chains and as a driver of dynamic pricing and personalized recommendation services, which aim to preempt user behavior and purchasing decisions. User information provides datasets for Amazon's AI services and feeds into its advertising services, through which companies can purchase advertising space on its platforms and sites. Amazon's access to customer information on a large-scale, as well as its capacity to formulate "exclusive" target audiences based upon this information, are promoted as features of these services.[61] Furthermore, Amazon "packages" information from user data, which is then sold to third-party advertisers and marketers.[62]

Drawing on analyses of free labor in the digital domain, the user might appear alongside the Foxconn worker and Amazon reviewer as those whose exploitation is implicated in the smart speaker and its ability to listen. However, such a framing of the user needs to be approached with caution, given that the "labor" of "free labor" has been subject to contention. As Michael Kaplan summarizes, there is a lack of consensus among media theorists as to whether user engagements within digital culture should count as labor, and the extent to which this labor is a source of industry profits.[63] Concerns have been raised that referring to users' activities in terms of labor risks creative false equivalences between relatively affluent consumers in North America, Europe, and Australia and the brutal working conditions of those working to manufacture and assemble digital devices.[64] Such critiques, however, tend to omit that some advocates of the free labor thesis have themselves made explicit that a direct comparison cannot be made to the alienated worker of industrial capitalism.[65] Furthermore, although they might not be equivocal, there remain important connections between the different modes of labor that underpin the functionality of digital technologies, in that "at every level contemporary technology is rooted in and running on the exploitation of human bodies."[66] As Nick Dyer-Witheford argues, the various types of labor implicated in digital technologies can be understood to drive one another: the toil of assembly line workers in China producing cheap computational commodities for the Global North creates the material basis of the platforms that generate what he terms "voluntary labor" for digital capital, which in turn propels further low-wage exploitation of electronics workers. The overall result is the increase in the power and wealth of major information corporations.[67]

Marxian critiques have sought to clarify what kind of labor is implicated in free labor, by questioning whether unpaid user activities captured through and by digital technologies can be considered generative of surplus value. Aspects of these debates recall those around reproductive labor's relationship to value, summarized in chapter 1. Christian Fuchs, for example, applies the labor theory of value to digital culture to argue that free digital labor is productive labor. Engagements with digital technologies and within the proprietary spaces of the internet are used to attract advertising clients, with the audience being "sold" by digital companies to advertisers as a commodity. To this end, Fuchs understands social media users as directly creating surplus value.[68] Responding to Fuchs, Dave Beech has raised various issues with this line of argument. Noting the apparent conflation of the user as both commodity and laborer, Beech challenges Fuchs's assertion that what is exchanged between digital companies and advertisers is a commodity (since advertisers do not purchase an audience but rent space) and the apparent assumption that advertising produces value: "advertising does not add value but only facilitates the realization of returns on value production through the sale of what is advertised."[69] Reminding us that "not all labour produces value," Beech concludes that while Fuchs is correct that digital companies such as Facebook cannot make money if users do not constantly engage with the platform, "it does not follow that these users are therefore productive laborers."[70]

I return to some of these debates regarding the relationship between users, digital platforms, and accumulation in chapter 5. For the moment, what I want to draw out of the free labor thesis is a more general acknowledgement that users and their activities contribute to the production and maintenance of corporate wealth in ways that go beyond the consumption of commodities. The smart speaker is a product that is purchased but it is also a means of data accumulation via questions, requests, and listening habits. In this regard, the listening performed by smart speakers is *extractive*.[71] Users' engagements and behaviors are subject to datafication, contributing the production of what Sandra Robinson describes as "data doppelgängers." A combination of information and statistical estimation, the data doppelgänger is a proximal object sprung from "continuously varied data sets from any number of databases" that feed into decision-making apparatus: from recommendation systems and targeted advertising to mortgage lender offers.[72] The data doppelgänger is not a mirror of an individual but rather is nonidentical, incomplete, and perpetually in formation: a product and constituent of profiling, prediction, and algorithmic apparatus that translate data into "insights." Indeed, as the Amazon technologies patent identifies (figure 6), within this digital economy, the user is not simply conflated with an individual identity, determined in accordance with various demographic parameters. Rather, the surveilled user is conceived of as multiple, changeable, and fluid, an aggregation of habits, rhythms, affects and drives.

It is the logic of extraction—which is to say, the accrual and privatization of data from user's engagements—that structures the "listening" of digital assistants and smart speakers. Their proximity to reproduction, domesticity, and femininity should not conceal these devices' connections to digital regimes of exploitation and monetization: "surface" functionality should not obscure their material "depth." Instead, there is a need to attend to the *coherence* between feminization, extractive listening, and the smart speaker's domestic domains.

## INCENTIVIZING SURVEILLANCE

In episode six of the second series of *Succession*, the revered satirical television comedy-drama centering on the Roy family and their media and hospitality conglomerate Waystar Royco, members of the corporate team attend Argestes, a business and tech conference, seeking to finalize their takeover of a rival media giant. The episode sees Tom (son-in-law of the company patriarch, Logan Roy) and Greg (Logan Roy's great-nephew) prepare for a presentation on behalf of ATN: Waystar Royco's right-wing news network. Tom is excited about unveiling the network's new audience-focused slogan: "ATN: we're listening." However, Greg warns him that "legal and comms" were not thrilled with the proposed slogan. It turns out that Waystar's set-top boxes have, in fact, been listening covertly to consumers. Greg informs Tom that the box's voice-activation mode and a "gray area in our data collection" means that "we're actually listening, sometimes we're listening quite aggressively." Faced with a potential media scandal, the slogan needs changing. "We hear you" is floated ("it's more 'couldn't help glimpse you changing' and less 'we put a spycam in your shower'"), as is "We're hear for you" (dismissed as "gibberish") before the pair finally settle on the reassuringly ambiguous "We hear for you."

This short storyline neatly illustrates the anxious cultural reception that smart speakers have received. Amazon's recording of user interactions has been well reported, while concerns about the implications the "listening capacity of these devices have for privacy and security have been repeatedly raised, often with good reason: in 2023, for example, Amazon was instructed to pay $25 million dollars to the US federal trade commission over child privacy violations after it was accused of failing to delete Alexa recordings at the request of parents.[73] Depictions of smart speakers and digital assistants as "Orwellian" or "Kafkaesque" are common, as are somewhat crude comparisons to totalitarian regimes of auditory surveillance.[74] These suspicious representations of smart speakers contrast sharply with Amazon's own depictions of these technologies, where the Echo is a technology of convenience and Alexa is a benign but intelligent helper.

The performance artist Lauren McCarthy offers a comparatively rich creative interrogation of the politics of "assistive" surveillance with LAUREN, a project that began in 2017 in which she tries to become the human version of Amazon's Alexa. McCarthy describes the project as follows: "For three days, I remotely watch over the person 24/7 and control all aspects of their home. I attempt to be better than an AI, because I can understand them as a person and anticipate their needs." Combining ocular and aural surveillance, McCarthy installed networked devices, including mics, sensors, cameras, switches, and door locks to create "the human intelligent smart home." LAUREN adjusts lighting and room temperature, queues music, keeps track of medication schedules, and gives advice. She responds to commands and requests but also seeks to preempt them. Doing so works to conceal LAUREN's presence—participants suggest there were times they forgot she was observing them—but could also result in friction. In a "testimonials" video, one participant suggests that LAUREN sometimes misjudges what they want "internally" by playing music or switching off all electronics to help them "wind down."[75] As well as foregrounding the human labor that lies behind algorithmic processes, McCarthy suggests that LAUREN is concerned with how the promise of convenience comes at the cost of privacy and control over our lives and homes: "We give access to all our data and live camera feeds—for what?"[76]

The extent to which current and prospective users of smart speakers share these concerns about privacy is unclear. According to a report by the business consulting and technology services company Capgemini Research Institute, around 52 percent of 12,000 survey respondents in the US, UK, France and Germany were concerned about digital assistants such as Alexa listening in the background to their private conversations, while 49 percent of respondents did not trust their digital assistant with the safety and security of their personal data. However, the report also found that adoption and use of digital assistants had increased between 2017–2019.[77] A small 2017 study by Lydia Manikonda et al. found that around 69 percent of 50 Echo home users had concerns about privacy. When users were asked about the device's always on feature (of which nearly 20 percent of participants were previously unaware), privacy concerns increased. Yet the study also recognizes that online reviews of Amazon's Echo are largely positive.[78] Given the popularity of smart speakers and the understandable anxieties about their ability to listen in on their users, the question remains as to how adoption of these devices is incentivized.

One apparent means of encouraging consumption of smart speakers is to keep the price low. Amazon's "super affordable" Echo Dot costs around £30 or $50—the equivalent of what CNET describes as "a few meals out"—and these devices are very often available for even less during sales and promotional deals.[79] Indeed, in July 2020, during a US House Judiciary Antitrust Subcommittee Hearing, Amazon's Jeff Bezos was asked whether the Echo Dot was sold below cost. Bezos responded that it was not sold below cost at "list price" but "sometimes when it's on promotion it may be below cost." It was put to Bezos by subcommittee member Jamie Raskin that Amazon sells its speakers below cost (which is to say, at a loss) so as to undercut competition and "own the smart home."[80]

While the cheapness of smart speakers removes some barriers to expanding adoption of these technologies, the feminization of the smart speaker and its digital assistant may itself serve to assuage anxieties about the device's surveillant capacities. The use of "Alexa" as the smart speaker's wake word is itself is revealing of how alleviating concerns about surveillance informed the product development process: Bezos was reportedly keen on using "Amazon" as the wake word, resulting in a disagreement with the design team who felt that users would be hesitant to engage with an explicitly corporate entity. And while Bezos wanted the device to produce multiple, distinct voices each associated with different uses or tasks, the design team decided to use a single feminized voice—allegedly based on the voice of US singer and voice actor Nina Rolle—since "female" voices were associated with attributes such as trustworthiness, warmth, and empathy.[81] These design decisions seemingly affirm Heather Woods's claim that equipping artificially intelligent digital assistants with a feminine persona rhetorically disarms users' hesitancies relating to intimate data exchange.[82] The humanizing and sexualizing of the of the surveillant device and the programmed persona of the digital assistant "recreate and reify gender codes attached to domesticity . . . to entice users and would-be users into 1) buying devices, 2) using them on a quotidian basis in increasingly intimate ways and 3) relinquishing control of their personal data for the privilege of interacting with these artificially intelligent virtual assistants."[83] From this perspective, feminization is not only a reflection of the kinds of work these technologies promise to do; nor, as spokespeople for Amazon have claimed, is it simply a reflection of a consumer preference for "pleasing" women's voices.[84] Rather, through a remaking of the gendered and racialized relations, smart speakers and their digital assistants are able to appear servient rather than surveillant. To recall the *Succession* plotline, they are intended to be suggestive of "we hear for you" rather than "we're listening."

The amiability of the smart speaker's feminized persona is coupled with the naturalization of orality and aurality as instinctive, intuitive, and paradoxically unmediated. Such framings of sound, speech, and listening have been readily critiqued, yet the perceived intimacy, immediacy, and universality of sound, speech, and listening nonetheless remains a powerful social fiction.[85] Thus, on its developer site, Amazon describes voice-based communication as:

> one of the most natural ways to use a computer; it requires no learning curve. . . . instead of having to make several swipes and clicks to play music, you can simply say, "Alexa, play the top songs in Seattle." . . . With conversational AI, voice-enabled devices like Amazon Echo are enabling the sort of magical interactions we've dreamed of for decades. Through a voice user interface (VUI), voice services like Alexa can communicate with people in ways that feel effortless, solve problems, and get smarter over time.[86]

As I have already implied with reference to users "yelling" at Alexa, these "magical" and "effortless" interactions often fall short: instructions are misheard, and commands need to be repeated. Yet, as these statements attest, oral/aural communication is presented as minimizing the necessary gaps created by mediation: clicks and swipes are replaced with simply saying what you want. The design background of the Echo similarly reflects an attempt to partially conceal communication's reliance on mediation. In contrast to early iterations of Apple's Siri, which entered the market in 2011 and required users to directly speak into the smartphone's microphone, the device that became the Echo was intended to be able to distinguish voice commands from across a noisy room, enabling users to simply speak out loud rather than directly to a technological object.[87]

If the naturalization of orality/aurality and concealment of its conditions of mediation helps facilitate the smart speakers' extractive listening, so too does its (contested) status as a music playback device. As is made clear by the "Sharing is caring" advert, with which this chapter began, the ability to respond appropriately to requests for music is presented as one of the Echo's primary functions. Bezos himself was reportedly resistant to emphasizing the smart speaker's ability to play music: the device was envisaged as a science fiction–inspired generalized computer that pushed at the boundaries of established technology; he did not want it to be reduced to such quotidian features.[88] Nonetheless, Echo devices are marketed as allowing listeners to quickly and easily access music via streaming services without having to engage with a visual or haptic interface. The relationship between smart speakers and music streaming services is characterized by reciprocity and correspondence. With regard to the former, smart speakers both drive and are reliant on music consumption. A 2018 joint report by the British Phonograph Industry (BPI) and the Entertainment Retailers Association (ERA) suggests that playing music is amongst the most important uses of smart speakers, while smart speakers will draw subscribers to music streaming services. Regarding the latter, smart speakers and streaming services incentivize particular modes of music consumption: namely, "causal" listening that occurs alongside and in relation to daily activities, routines, and practices. In the same BPI-ERA report, it is noted that it is common for users of Amazon's smart speaker devices to issue commands like, "Alexa, play music," deferring decisions about artist and genre choices to the platform's algorithm.[89] Such requests cohere with the rise of the playlist as an organizational format within streaming services, with one song smoothly blending into the next.[90]

Despite Bezos's reluctant about its use for music consumption, Amazon is well placed to capitalize on the relationship between its smart speaker and streaming services, given that it owns both the hardware and the platform. The company's own streaming service, Amazon Music, is heavily promoted to current and prospective Echo users through reduced subscription fees and "bundle" offers. Like other music streaming services, Amazon's music platform is organized around genre, mood, and utility, and places emphasis on music's functionality and affectivity, as well as the ability to access new music and favorite songs. The "mood and activities" playlist categories include "family and fun," "energetic and upbeat," and "relaxed and laid back"; while playlists for studying, working, and reading include "classical focus," "jazz focus," and "gospel focus." Mediated by these categories, music is construed as a naturalized and continuous accompaniment to everyday life, as well as an ideal "technology of self," enabling us to reflect and refine through music our physical and psychic states.[91] However, as Eric Drott astutely observes, "what makes music so powerful as a 'technology of the self' . . . is what allows streaming platforms to repurpose it as an equally powerful technology of surveillance."[92] Music's long history of reproductive utility, its proximity to subjectification, and capacity to inconspicuously accompany the most intimate of moments in our day to day lives make it an ideal means of accessing user information and attention. Consequently, the smart speaker's ability to access and play music aligns with both its presentation as a household helper and its extractive logics.

Although the feminization and naturalization of the listening smart speaker may help assuage anxieties about auditory surveillance, it would be inaccurate to conclude from this that naïve users are simply duped into allowing access to their data. Rather, the gendered constitution of these devices also helps to reconfigure the negative connotations of auditory surveillance by making it a *service*. The Echo's capacities for extractive listening are presented as something that primarily benefits the user. As McCarthy's LAUREN makes explicit, intimate details are key to the device—and, by extension, Amazon—"knowing" the user, enabling, for example, the delivery of the "right" music choices, even when provided with vague instructions and requests. The ability of Alexa to "remember" previous interactions is articulated as a feature to current and prospective customers, while the "always on" status of the smart speaker helps to make its operations frictionless, enabling it to react whenever users "just ask." As Emily West argues, "Fundamentally, Amazon offers to serve us by *knowing* us," and "Amazon's tools and techniques of surveillance create tremendous *intimacy* between consumer and brand achieved through the sensations of being seen, heard, and known."[93] The collection and retention of information and the preemption of needs produces and maintains the connection between user and device, and with it, customer and Amazon. The accrual of data is made palatable—or even desirable—by a device that promises to use its ability to listen to us to care for us.

## ECHOES OF AUSTERITY

In 2016, when the Echo was first made available in the UK, Britain was six years into a period of austerity initiated in 2010 by David Cameron's coalition government. Presented as a necessary course of action following the 2007–2009 Great Recession and the 2008 financial crisis, public expenditure was significantly reduced in the name of ameliorating the government budget deficit. Social housing, education, libraries, arts and culture, and the maintenance of roads and infrastructure endured budget cuts, which were rationalized by the need of the country to "take responsibility" for its debt. The adult social care sector, likewise, was severely impacted by the rescindment of public spending in the face of growing demand and costs. Between 2010–2019, English local authorities (who have a statutory responsibility for social care provision cut social care funding by £7.7 billion in real terms.[94] Meanwhile, it was anticipated that the over 85s—the age group most likely to require social care—would rise from 1.6 million to 1.8 million between 2015 and 2020.[95]

Given these challenges, it is unsurprising that many local authorities have turned to technology to try and alleviate the disjuncture between provision and demand. In 2018, Hampshire County Council became the first local authority to work with Amazon in a trial in which fifty adult social care service users were provided with customized Echo speakers.[96] The trial was part of a wider council scheme, arising from a partnership with the consortium PA Argenti, whose "innovative approach to technology-enabled care" is reported to have resulted in multimillion pound savings for councils.[97] Amazon's smart speakers have since been trialed by various other English local authorities, including Oxfordshire, Richmond, Cornwall, and Norfolk, as well as being increasingly common in residential care homes: a September 2019 survey of 2,611 care-home workers found that one in four care staff reported their workplaces using Alexa or Echo to assist or entertain residents.[98] In a BBC report on the Hampshire County Council trial, one participating couple, Claire and Eric Williams, describe the ways that the Echo smart speaker has helped them. Eric says he is able to go out without worrying as Claire, whose mobility is restricted by multiple sclerosis, can use the device to call him; Claire demonstrates how the Echo enables her to easily play music (specifically Barry Manilow) and operate the lights in her home. Eric says that while the speaker can't address physical needs, a carer coming in costs money: a smart speaker, by comparison, only needs to be paid for once. Claire adds, "You need other stuff other than physical stuff," detailing how the device helps meet her emotional and mental needs.[99]

As the Williams's testimony suggests, the use of smart speakers as assistive technologies can offer benefits to users with disabilities. Amazon has itself recurrently emphasized Echo and Alexa's assistive capacities in relation to elderly and disabled users. In addition to the "Sharing is caring" advert, in which the Echo is shown as helping a grandson to look after grandpa and helping grandpa to look after himself, Amazon have produced an audio-described advert in collaboration with the UK charity Royal National Institute of Blind People (RNIB), depicting a blind user's morning routine: she asks Alexa the time of day and what the weather is like, before leaving the house with her guide dog. And in November 2020—during the COVID-19 pandemic—Amazon released Care Hub (rebranded as Alexa Together in 2021): a set of Alexa features for US users intended "to help aging customers maintain independence and provide assurance to family members."[100] The smart speaker's extractive listening and surveillant capacities are, in this context, understood to provide peace of mind: presenting the family as the primary means of care for older people, Care Hub notifies a "supporting" family member when an "aging loved one" interacts with Alexa, as well as when no activity has been detected after a specified time period.

Amazon's Echo and Alexa, then, can potentially support elderly and disabled users to live well by helping meet their emotional and mental needs and aiding them with their daily routines, while also providing reassurance to members of their support network. Yet the adoption of smart speakers as assistive technologies in care settings has been partly driven by their capacity to reduce the cost of social care. A council report at the beginning of the Hampshire trial highlights how assistive technologies, including Echo speakers, have enabled the council to make "substantive" savings: in the context of an aging population and finances becoming ever more stretched, "making greater use of care technology was identified as a way of both saving money and providing valuable support to residents."[101] Similarly, in a report on a small trial by Oxfordshire County Council with Amazon's Echo, during which Amazon provided the devices for free, installed broadband in participant's homes, and delivered training for staff and service users, it is suggested that the speakers could "allow some care resources to be freed up and directed to those in greatest need of personal visits."[102] Kate Terroni, as the council's Director of Adult Social Care, says of the Oxfordshire trial, "While many job sectors are losing out to advances in technology this is not the case in the care sector. Technology can enhance people's lives rather than replacing the vital role care workers provide. You can never replace the warmth and empathy of human contact, meaning a career in the care industry is future proof."[103] However, a notably different line was adopted in a news item for Community Care that reported on how councils have worked with Amazon to explore how its technologies "can deliver more personalized care to service users and relieve pressure on a stretched social care workforce": here, Terroni emphasises the challenges of recruiting and retaining social care staff in a county that contains one of the most expensive UK cities to live in. Consequently, "reducing demand on the workforce" is described as a core motivation behind the Oxfordshire County Council's trial scheme.[104]

Given the timing of both the Oxfordshire and Hampshire trials and the emphasis on making "savings," austerity can be recognized as a factor driving the adoption of smart speakers: for English country councils, Amazon's Echo and Alexa are a stop gap for the holes left by underinvestment and an overstretched workforce.[105] Indeed, these trials exemplify what I want to name as *techno-austerity*, whereby technology—broadly defined—is identified as enabling services to be appropriately reconfigured around the demands of austerity and an obligation to do more with less. In this regard, techno-austerity is a distinct yet related component of a broader trend, in which, as discussed in the introduction, sound, music and their related technologies appear as a cheap fix for the capitalist crisis of reproduction, with low-cost consumer sound technologies positioned as surrogates for "high-cost," labor-intensive services. What distinguishes techno-austerity, however, is that technological automation and assistance, and the accompanying language of efficiency and "innovation" are presented as a necessary means of cutting costs to services as part of a broader incentive to reduce social spending.[106] To this end, the smart speaker, in direct contrast to the futuristic fantasies of ease, convenience, and digital abundance to which it is frequently attached, is an emblem of scarcity, shortage, and incapacity.

Although techno-austerity is often accompanied by an emphasis on improving experiences of service users, it has had a detrimental impact on those in need of care. In May 2018, a complaint was upheld against

Devon County Council after a service user's supportive living arrangements were changed so that nighttime care support was withdrawn and replaced with a voice-controlled assistive technology. The service user, a resident in supported living accommodation with a hearing impairment and "poor communication skills," experienced "significant distress" because of the changes. The service user's parents allege that when they raised concerns about the changes to his plan and the suitability of the assistive technology, they were told by the service user's social worker that the council were "obliged" to try it as a more cost-effective way of providing support. The care provider had also raised concerns with the council, flagging that changes to the service user's care were causing distress and had resulted in multiple erroneous calls to the emergency services. The Local Government and Social Care Ombudsman report into the case concludes:

> The Council says it looks at ways of promoting independence. However, that has to be balanced against risk and it appears that in this case, the balance was tipped too far. There is at least an appearance that the Council intended to trial the AT [assistive technology] whatever Mr and Mrs S said during [the service user's] assessment. It is difficult to see how the Council otherwise reconciled the assessment that [the service user] was unable to respond appropriately to emergencies or let others know if he was in pain, with the introduction of voice-activated technology for someone who was hearing impaired and had poor communication skills.[107]

While the manufacturer of the voice-controlled assistive technology is not named in this case—there is no suggestion that it was an Amazon device that was used to replace the service user's nighttime care—it illustrates one of the possible outcomes of techno-austerity, as well as being an important reminder of the ableism inherent in the assumption that that anyone can "just ask."[108] Such cases also illustrate how techno-austerity is accompanied by significant tensions between the investments and priorities of national and local government, care providers, and service users. For instance, the aspiration of "independence" referred to by Devon County Council—which is often attached to assistive technologies—is by no means neutral, nor is its definition uncontested: what constitutes independence for policymakers and care providers (e.g., a lack of reliance on a service, leading to cost-reduction) is not always the same as what constitutes independence for those accessing care services (e.g., the ability to take risks, to engage in activities previously enjoyed).[109] Independence that is imposed due to the demands of austerity and is driven by cost reduction differs substantively from independence as a state that is produced by facilitating autonomy and access for elder people and people with disabilities.

The adoption of Amazon's smart speakers in England's adult social care settings reveals the potential beneficiaries of both austerity and techno-austerity, insofar as partnerships with local authorities offer technology companies access to data, contracts (often funded by government spending), and potentially lucrative ideas, as well as opportunities to expand their consumer base. As Caleb Johnston and Geraldine Pratt argue, despite tech companies often framing their collaboration with local authorities through the language of largesse (i.e, offering their technology services for free), local authorities are ultimately paying for product development: "Local authorities come up with the questions, write grants to seed projects, write more grants to trial the projects, and absorb the risk of implementation. The results of these trials are then used by companies to pitch their products to other local authorities and . . . commissioning officers with the NHS. Global health markets are within their sights."[110] In addition, local authorities facilitate access to the personal information of residents and service users, with council schemes involving Amazon services requiring participants to sign up for an account with the company and accept their data agreements.[111] The presentation of the smart speaker as a feminized "caring" device thus enables and is reinforced by its transition into the social care sector, which, in turn, provides new opportunities for expansion, extraction, and accumulation. At the same time, and as is well evidenced by Amazon's Care Hub service, the "care" offered by the smart speaker is constrained by a limited and well-established imaginary that is preoccupied with keeping people in their own homes for as long as possible and out of expensive alternatives such as long-term care facilities.[112] Contrary to the rhetoric of innovation, transformation, and "disruption" espoused by Bezos and other tech leaders, techno-austerity forecloses questions about the norms of care and care work so that automation appears as the sole alternative to a precarious, poorly paid, and time-stretched workforce.

## CONCLUSION: SWITCHING OFF

Built on a domestic imaginary in which orality are central to technological assistance, the smart speaker and its digital assistant present voice-based communication as simultaneously effortless, intuitive, and futuristic. This construct of *oral*ity, however, is accompanied by a gendered and racialized construct of *aural*ity, allowing the device to appear as an open, attentive yet selective listener that is both affectively attuned and emotionless. Powered by a substantive, transnational, and environmentally deleterious network of resources and labor, the smart speaker's listening facilitates and normalizes relations of surveillance and extraction that are central to its two-sided functionality, enabling it to respond effectively *to* the user, as well as to accumulate data *about* the user. These users may be brought into relation with the listening smart speaker by purchasing the device themselves, or via a state intermediary such as a local authority. Consequently, these devices are expressive of the intensified marketization of both the personal and the public.

To recognize the smart speaker as a listening device enables us to acknowledge the coherence between its gendered presentation and Amazon's data-economic logics. However, this account of the smart speaker also presents challenges the articulation of listening as a virtuous sensory relation within sound studies and feminist theory. The recurrent advocacy for attuned, open, and attentive listening is contraposed by the Echo as a reminder of listening's proximity to feminized work and practices of auditory surveillance. The valorization of listening as a site of ethical action without attending to its exploitation risks reinforcing, rather than undermining, its uneven distribution. There are, after all, situations where good, open, and attentive listening may in fact be harmful and exploitative and where a *refusal* to listen may be necessary. A feminist ethics or politics of listening, then, needs to allow for "switching off."

It is the incapacity to switch off that is integral to the smart speaker's extractive listening: to engage with Amazon's Echo or Alexa is to contribute to the production of data, and opportunities to effectively "withdraw" this "free labor" are limited. However, this is not to suggest that Amazon's business model is resistant to any disruption: despite the company's noted hostility to trade unions, there have been strikes by European logistics workers, work stoppages in US warehouses led by Somali workers demanding Islamic prayer rights, walkouts by tech workers demanding that the company address climate change and stop assisting US federal immigration enforcement, and protests by warehouse workers over health and safety concerns during the COVID-19 pandemic. The concessions gained and enduring effects of these actions have been modest. However, as the corporation's power and accumulation of wealth grow, so does discontent about its practices. If, as Donna Haraway and Lauren McCarthy encourage us to recognize, the listening smart speaker is reliant on human labor, it is also vulnerable to its disruptions.

# 3 "Your Womb, the Perfect Classroom"

## PRENATAL SOUND SYSTEMS AND UTERINE AUDIOPHILIA

In December 2015, the Catalan startup company Babypod / Music in Baby S.L.U., in collaboration with singer and 2009 Eurovision entrant Soraya Arnelas, held the first "concert for fetuses."[1] Attended by ten future parents, the performance largely consisted of acoustic renditions of Christmas songs and carols. Video documentation from the concert shows the singer performing with piano and guitar, while ultrasound images are projected behind her. Later in the video, visibly pregnant white women surround Soraya on stage, clapping and swaying in time to the music. Many are holding a smartphone in their hand attached to a pink cable.[2] This cable belongs to the Babypod intravaginal speaker, a device developed in collaboration with the assisted reproduction center Institut Marquès, which is used to transmit the concert to the fetuses. Institut Marquès has an interest in the effects of music at "the beginning of life": in addition to the "concert for fetuses," they have held a "concert for embryos" featuring Spanish singer-songwriters Antonio Orozco and Álex Ubago and Irish musician Sharon Corr as part of their advertised work on the beneficial effects of music on in vitro fertilization outcomes.[3]

In allowing music to be played in utero, Babypod claims to stimulate vocalization in babies before birth, helping to develop their communication skills in the womb. The device consists of a small pink silicone speaker capable of broadcasting sound up to fifty-four decibels and can be attached via mini jack cable to a phone. With the tagline "music is life" and a price of €150, the Babypod speaker is designed to be used from the sixteenth week of pregnancy: by this stage, fetuses reportedly perceive sounds as distorted whispers, and when played music, they respond with tongue and mouth movements. These are, in turn, equated with attempts at vocalization and thus with communication development and language learning. The Babypod website states:

> Music stimulates major brain structures in human beings, it improves their neurological development. Therefore, paediatricians recommend that children grow up in environments with many stimuli, including music; What if you could start now? Thanks to Babypod children can begin to enjoy the pleasure of music before birth. . . . Music stimulates comunication [*sic*] and language, the form of expression and the coordination of movements. . . . Thanks to Babypod children begin to vocalize in the womb.[4]

With patents for prenatal audio apparatus dating back to the 1980s, Babypod is one of a range of prenatal sound systems available to expectant parents.[5] These sound technologies promise multiple benefits: from soothing away stress to increasing the efficiency of sonograms. However, one common point of emphasis is their capacity to accelerate fetal learning through sonic stimulus. There is, for example, the BabyPlus Prenatal Education System, the tagline for which figures "your womb" as "the perfect classroom" (this tagline has been more recently updated to "the right sounds at the right time"). Founded by the US inventor Brent Logan and distributed internationally—the company website lists retailers in nineteen different countries, including China, Singapore, Hong Kong, Malaysia, Brazil, Mexico and Nigeria—BabyPlus consists of a belt speaker system programmed with a "curriculum" of auditory stimulation lessons "designed to spark cognitive development and a unique bonding experience between mom and baby during pregnancy."[6] This program of environmental enrichment is intended to teach what BabyPlus describes as "prenatal children" to discriminate between different rhythmic patterns and, in so doing, create a "stronger cognitive foundation" for future learning.[7] In a BabyPlus marketing video, East Asian BabyPlus infants are depicted performing "cuteness"—sitting at the piano, playing with toys, laughing and dancing—in contrast to a noisy, screaming, unhappy infant that has not experienced prenatal enrichment. The voiceover informs the viewer that playing rhythmic "natural" sounds akin to a heartbeat via the BabyPlus device helps ensure that babies develop better sleeping patterns and an increased capacity to self-soothe and are more relaxed, alert, and interactive, while in later life, BabyPlus children achieve earlier developmental milestones, improved school readiness, enhanced intellectual abilities, and greater creativity.[8]

Prospective motherhood has become a key site of capitalist consumption: there are a plethora of commodities, including books, websites, CDs, playlists, education classes, cosmetics, supplements, services, and apps, available to help track, understand, and enhance a pregnancy. However, this chapter's focus on prenatal sound systems does not stem from their widespread popularity. Unlike some of the other technologies addressed in this book, prenatal sound systems are not the product of or adjacent to large, monopolistic companies (e.g., Amazon, Universal Music Group): their consumer market appears limited, and press coverage of these devices tends to focus on the novelty factor. Despite this, I contend that these technologies are worthy of critical attention. Where feminist critiques of pregnancy and technology have often focused on visual modes of representation and mediation, prenatal sound systems make apparent the significance of the auditory for cultural, economic, and political figurations of gestational labor and fetal life.

As with other technologies discussed in this book, prenatal sound systems operate and circulate in ways that draw together multiple notions of reproduction. This includes the "reproduction" of assisted reproduction. Assisted reproduction is typically thought of as medical procedures and practices that address issues related to fertility and conception, such as surrogacy, in vitro fertilization, and gamete intrafallopian transfer. While some prenatal sound systems have been developed in proximity to these procedures—the Babypod speaker, for example, is said to have been developed in relation to Institut Marquès's research into music's impact on in vitro fertilization[9]—they are ultimately nonmedical "well-being" devices that promise to provide environmental enrichment. Nonetheless, there remain points of continuity that enable prenatal sound systems to be productively framed in terms of assisted reproduction: both pertain to technological interventions that assist, facilitate, and "optimize" gestation, appearing at the intersection of the biological and social. Prenatal sound systems, like assisted reproduction more broadly, concern the "making" of babies and parents, raising political, ethical, and ontological questions about pregnancy, childrearing, and kinship. Charis Thompson notes that assisted reproduction and its associated technologies operate in relation to multiple "relevant kinds of time": there is, for example, the repetitive time of medical and treatment cycles, the bureaucratic time of the working day into which appointments must be fitted, and the linear, nonrepetitive time of "biological clocks." Prenatal sound systems likewise encompass multiple temporalities. This includes rhythmic or musical time, the repetitive time produced by using these devices on a regular basis, and the linear time of fetal and cognitive development that these technologies aspire to accelerate. As technological corollaries of human capital, they also involve a folding of time, whereby the future of the child is projected onto the preborn and their uterine environment. Furthermore, just as assisted reproduction attests to the intertwining of reproduction's numerous literal and metaphorical meanings emerging from the intersections of technology, economics, biology, and society,[10] prenatal sound systems enact and connect multiple reproductions. They bring sound reproduction technologies into relation with processes, practices, and institutions of biological reproduction, and in doing so, they also function as technologies of social reproduction. As devices intended to secure a competitive future, these devices affirm uneven and inequitable configurations of risk and reward that accompany financialized life.

This chapter approaches prenatal sound systems in accordance with the logics of their design, the (dubious) claims that are made regarding their efficacy, and their auditory and reproductive imaginaries, asking what kinds of futures, temporalities, and "listeners" these technologies seek to produce. Taking the Babypod and BabyPlus devices as exemplary, I argue that prenatal sound systems make audible a particular figuration of pregnancy and gestational labor that combines divergent notions of responsibility and passivity. As sonic surrogates that automate and enhance maternal-familial "investment" in the future-child, prenatal sound systems enact a biopolitics of impressibility. However, where the fetus is configured in terms of plasticity, transformability, and the capacity to listen, prenatal sound systems figure the pregnant body as a resonant container. Consequently, these devices exemplify what I refer to as *uterine audiophilia*. Highlighting the coherence and tensions between common depictions of sonic experience in terms of uterine spatiality and the fidelity discourse apparent in prenatal sound systems' design and marketing materials, I contend that uterine audiophilia serves to disappear the broader matrix of reproductive relations.

## SOUND INVESTMENTS

In his 1978–79 lecture series, "The Birth of Biopolitics," the philosopher Michel Foucault identifies the notion of human capital as a distinctive attribute of US neoliberalism, representing both "the extension of economic analysis into a previously unexplored domain" and "the possibility of giving a strictly economic interpretation to a domain previously thought to be non-economic."[11] Constituted by a person's skills, qualities, knowledge, and capacities, human capital theory recognizes these attributes as an "asset" that can generate a "return": be that a wage, credit, or another form of income. Consequently, Foucault describes human capital theory as concerned with the production of the individual as an "abilities machine." These "abilities" can be inherited via genetics and "good breeding" or acquired through education, upbringing, and experience. It is therefore the mother that is key to producing human capital by "investing" in the child, both through their reproductive choices (i.e., how many children they have and with whom) and, perhaps more importantly, through their caring practices:

> We know that the number of hours a mother spends with her child, even when it is still in the cradle, will be very important for the formation of an abilities machine, or for the formation of a human capital. . . . Time spent, care given, as well as the parents' education—because we know quite precisely that for an equal time spent with their children, more educated parents will form a higher human capital than parents with less education—in short, the set of cultural stimuli received by the child, will all contribute to the formation of those elements that can make up a human capital.[12]

Not only are women "the invisible infrastructure sustaining a world of putatively self-investing human capitals";[13] they are also the first investor. To this end, Foucault proposes that a child can be subject to "environmental analysis" such that aspects of their family life can be measured in terms of "the possibilities of investment."[14]

Originating with the Chicago School economist Theodore Shultz in the early 1960s, and further developed by his colleague Gary Becker, human capital has, in the words of Michelle Murphy, been "a world-shaping product" that has directed private, philanthropic, and government expenditure.[15] For Shultz, engagement with education and healthcare, as well as the development of skills and knowledge through leisure time are not simply acts of consumption or reproduction but rather investments in the subject. Contra the classical economic treatment of labor as "a capacity to do manual work with little knowledge or skill, a capacity with which . . . laborers are endowed equally," Shultz describes a scenario in which workers "have become capitalists . . . from the acquisition of knowledge and skill that have economic value."[16] Consequently, human capital theory disavows the fundamental antagonism between labor and capital: if all workers and would-be workers are competing capitalists, then exploitation disappears from the frame. As a neo-Keynesian, Shultz recognized investments in human capital as having benefits that extend beyond the individual and contribute to social prosperity. As a result, Schutz advocated for public education and other forms of government spending. By comparison, Schultz's neoliberal colleague Becker proposed that education and other investments in human capital should be primarily funded by the individual or the private household, be that through family inheritance or through credit, loans, and debt. Writing in the early 1980s, Becker posits that publicly funded education should be opposed on similar grounds to progressive taxation: in both cases, inequality is raised "because families reduce their investment in their descendants."[17]

Prenatal sound systems are a technological corollary to the financialized logic of human capital and its (oft-overlooked) maternal-familial imperatives. These devices enable prospective parents to develop a future "abilities machine" while in utero in the hope of securing a successful future. The BabyPlus website itself makes explicit reference to human capital: a section promoting "early investment," which here is understood as providing environmental "enrichment" for prenatal and early years infants, is accompanied by a basic graph credited to the economist James Heckman. A former colleague of Becker and fellow Chicago School member, Heckman is known for his research on the economic efficacy of childhood education. For Heckman, inequality primarily concerns opportunities for skill formation. Children who have early opportunities to develop academic and social skills are more likely to be economically successful in later life, while those born into "disadvantage" are comparatively lacking in early enrichment. Exemplifying the persistent interest of neoliberal economics in the family, Heckman proposes that children from disadvantaged family environments "are more likely to commit crime, have out-of-wedlock births and drop out of school." Early interventions into children's skill formation "can reverse some of the harm of disadvantage and have a high economic return. They benefit not only the children themselves, but also their children, as well as society at large."[18] Consequently, Heckman argues for investment in children from disadvantaged backgrounds with a view to increasing productivity and reducing risks to security at a population level. Although "most Americans are reluctant to intervene in the early years and prefer to respect the sanctity of the family," the earlier interventions can be made into a child's life, the better.[19]

This thesis informs the "Heckman curve" (figure 7), which features on the BabyPlus website.[20] Appearing under the heading "Early childhood development is a smart investment" and the subheading "the earlier the investment, the greater the return," the graph plots the rate of return to investment in human capital against the life stage at which the investment takes place. "Prenatal programs" are shown as giving the greatest rate of return, and post-school job training the lowest. Heckman's graph thus exemplifies how economics is used as a means of explaining, rationalizing, and justifying ostensibly social initiatives concerning childrearing.[21]

If human capital is partly produced through engagement with what Foucault refers to as "cultural stimuli," then prenatal sound systems seek to deploy recorded sound and music to this end. Their treatment of auditory materials as a means of developing the future-child resonates with a broader interest in music's social, economic, and communicative benefits. In music education, there have been various market and state interventions that seek to attach the acquisition of musical skills in childhood to

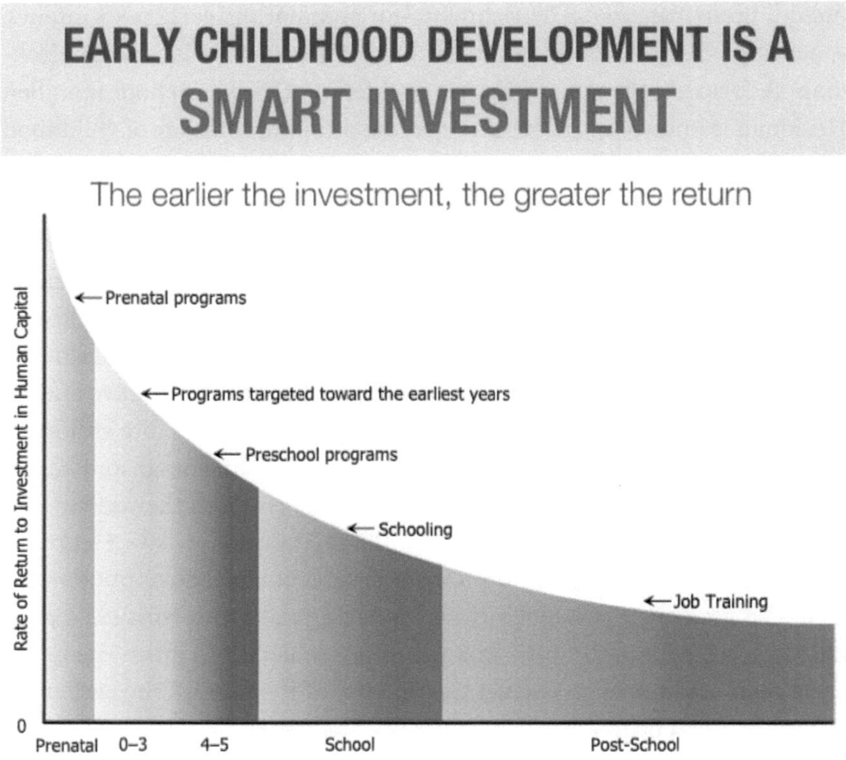


*Figure 7.* The Heckman curve as it appears on the BabyPlus website. From heckmancurve.org.

future success in adulthood. Opportunities such as instrumental tuition are thought to offer an appreciation in human capital, in the form of improved cognitive development and academic attainment, as well as improved musical skills. Both researchers and music education advocates have emphasized the need for early and sustained pedagogic opportunity as key to realizing these apparent extramusical benefits to music education.[22] This is not to diminish the variety of motivations and beliefs that accompany children's musical activities. As Beatriz Ilari has identified in her comparative international study of middle-class children's organized musical activities, the parental expectations of music education can vary widely: from being understood as a source of enjoyment and pleasure to being approached as a means of keeping children occupied while caregivers are busy.[23] Nevertheless, the perceived potential for music education to enhance children's future academic, social, and economic prospects has been a recurrent theme and rationale among policymakers and education providers.

Where the purported benefits of music for human capital are typically attached to developing skills in music performance, in the case of Babypod and BabyPlus, it is through other, specific forms of fetal "musicking" that sound and music become technologies of speculation. In the marketing materials for devices such as Babypod and BabyPlus, sound is posited as a means of connecting with and accelerating the development of the fetus. Aurality is frequently presented as the primary sense that the fetus encounters in the in utero and ex utero environment, and thus sound becomes the ideal means of engaging and "educating" the fetus, accelerating the child's capacity for vocalization and language acquisition. These devices are underpinned by an understanding that providing opportunities for "learning" during key morphogenic stages leads to greater intelligence in the future. According to BabyPlus:

> Today's mom knows that when it comes to providing an enriched environment for learning and intellectual development, earlier is truly better. Although your child will continue to develop cognitively into adulthood, it is through an early enriched environment that specific pathways can fine tune your child's memory, language, cognitive development and lifelong critical thinking.[24]

Prenatal sound systems present learning as that which requires intervention both through consumption (i.e., the purchasing of the device and its learning program) and through the adoption of technologically mediated childrearing practices by the pregnant person (BabyPlus recommend using their speaker system for an hour a day during pregnancy). Their emergence as a small but distinct subsection of the pregnancy industry is largely illustrative of parenting norms predominant amongst the middle-classes in affluent countries, for whom mothering has moved significantly beyond the provision of food, love, and shelter, and toward what Annette Lareau describes as "concerted cultivation"[25]—that is, an increasingly interventionist approach to child development whereby parents seek to hone and refine their children's abilities and "talents." In alignment with this logic, prenatal sound systems use sound to "enrich" the uterine environment to (in the words of BabyPlus) "fine tune" the prenatal child's cognitive skills.

While sound is presented as the ideal medium for enriching the fetal environment, the types of sonic stimuli that are considered most effective in enabling and improving fetal learning varies between different prenatal sound systems. The BabyPlus device emphasizes rhythm and "natural" maternal sounds. Their website states: "Our sounds are developmentally 'relative' to the natural prenatal environment and introduce simple patterns to the prenatal child in the only language he/she understands—the maternal heartbeat."[26] The fetus is understood to compare and contrast the sounds of the pregnant person's heartbeat with those generated by the device. These sounds become incrementally more complex in tone and rhythm over the course of the BabyPlus curriculum. Indeed, in the product's Frequently Asked Questions, BabyPlus suggest this program consisting of simple sound patterns and incremental change is superior to speaking or playing music to a fetus: "Researchers note that to promote pattern recognition, such basic stimuli must be very simple and repetitive. . . . Even nursery tunes designed for use as music for babies in womb—let alone Mozart—are simply too complex to register. . . . All babies are exposed to speech and prenatal music, but these pass by them as white noise, with no longer-term cognitive impact."[27] BabyPlus thus uses its sonic curriculum to differentiate itself from competitors: where the fetus experiences voices or music as white noise, the similarity of Baby-Plus's sound patterns to the maternal heartbeat means that they can be "recognized" by the fetal listener.

In direct contrast to BabyPlus, Babypod presents music as the ideal sonic stimulus for enriching the fetal environment, with the company noting music's neurological, autonomic, and emotional effects. Music is framed as an "ancestral communication" between humans, which precedes the development of spoken language.[28] Indeed, Babypod's founder and director of Institut Marquès, Marisa López-Teijón, claims that music's connection to emotion is central to its efficacy. When asked in an interview whether music needed lyrics or vocals to generate a fetal attempts at vocalization, López-Teijón states:

> It's rhythm and melody. And it's not just rhythm, because when we put on African [*sic*] drums, they don't respond. If we just put on melody, it doesn't work either. The difference between saying something through speech or through a song is that music activates emotions. So they're different circuits. For example, if we speak normally, babies respond much less than if we speak with musicality. With musicality there is a much bigger response.[29]

Such claims complement long-standing, popular imaginations of "the power of music": its ability to impress upon and connect bodies and minds, to modulate feeling and to reshape listener's perspectives and capacities, and to function as a universal mode of expression. These tropes, combined with music's dissemination as a cheap cultural resource, mean that it has frequently been promoted as a tool of affective modulation during pregnancy: it has been framed as a means of soothing stress and anxiety and as a method of pain relief during labour. It has even been proposed that regularly listening to music during the third trimester of pregnancy can be protective against symptoms of postnatal depression: a 2018 preliminary cohort study by Daisy Fancourt and Rosie Perkins suggests that there is a correlation between the quantity of musical listening and maternal mental well-being up to three months post-birth. As a result, the researchers suggest that music listening could be an "attractive intervention," particularly given that it is "readily available to people from all echelons of society. . . . It can be carried out in a range of contexts so is not restricted to particular places or times. It is also inexpensive: indeed, the majority of women in Western countries have access to recorded music already."[30] This study's concluding remarks reaffirm the connections between recorded music's apparent ubiquity, its cheapness, and its (actual or potential) use in healthcare settings.

Given the persistence of the (widely discredited) Mozart effect hypothesis, which postulates that exposure to Mozart and other types of classical music can improve intellectual development and performance, it is unsurprising that Babypod cite Mozart's Serenade K525 as the most effective musical stimulus.[31] Babypod report that Institut Marquès has tested fetal responses to a variety of sound and music samples, by measuring the number of fetuses that move their mouth and tongue to a given stimulation.[32] In addition to Mozart, compositions by Bach, Beethoven, and Prokofiev are listed as particularly effective at stimulating fetuses.

López-Teijón suggests that "traditional" music is also particularly effective, while fetuses "don't like" modern music: "It doesn't stimulate communication."[33] Despite these comments, Babypod's published playlist is more heterogenous than a straightforward comparison to the Mozart effect might suggest: it implies, for example, that Mozart's Piano Concerto No. 23 in A Major, K488 was less effective at generating fetal movements by comparison to Queen's "Bohemian Rhapsody" and The Village People's "YMCA."[34]

Prenatal sound systems are primarily intended to work on the preborn as future-child. In so doing, these devices simultaneously affirm, automate, and expand maternal-familial responsibility for the child and their life chances: they offer reassurance to prospective users that through purchasing and deploying these technologies they are fulfilling their role as "good parents." Babypod, for example, suggest that their device enables "the pregnant mother to ensure fetal well-being," while BabyPlus urges its users, "Give your baby the BEST start at a time when it matters most." Here, giving baby "the best" is extended to include using sound reproduction to produce the "optimum" auditory environment for the future-child and for their development of, or as, human capital. In so doing, mothers can ensure that they are meeting their responsibilities by enabling the future-child to excel in their learning and development, which is evidenced by meeting developmental milestones earlier than their peers.

In affirming the pregnant person's maternal-familial responsibility, prenatal sound systems also affirm the child as a figure of maternal-familial *investment*. Who gets to be a child has always been partial and selective: it is a status predicated upon particular racial, gendered, classed, and sexual norms and indebted to Eurocentric, bourgeois notions of the nuclear family (see chapter 4). However, capitalism's mutations mean that previous figurations of the child—as morally, economically and spatially distinct from the realm of the adult—have been destabilized. In his work on the child in popular music, Paul Rekret argues that neoliberalism and finance capitalism's reconstitution of the boundaries between private and the public, home and factory, work and leisure, productive and reproductive labor threatens not only the spaces of childhood but also its temporal coordinates. As finance capitalism enters the household and every element of life becomes a potentially tradable security, the common association of the child with leisure and enjoyment becomes increasingly precarious.[35]

Childhood, meanwhile, is increasingly structured around preparing for a future that is necessarily competitive. The social and ontological insecurity resulting from the redistribution of public and corporate responsibility for the social wage onto individuals, families and civic institutions is manifest in anxieties about the future, which are "channeled in and through concerns about children and the nature of childhood."[36] Writing in the context of waning US-European presumptions of hegemony and rule, Cindi Katz suggests that uncertainties about political-economic, geopolitical, and environmental futures are relayed into securing children's futures and producing "perfect" childhoods. Katz notes that the figuration of the child as a family investment long precedes the current era of reproductive precarity: prior to industrialization, for example, children were understood as an asset that could contribute to the household's regimes of production and reproduction, helping secure the economic future of the family. However, as the conditions and availability of work become increasingly diminished, it is common for individuals and families to scramble to ensure that their children "make it": hence the rise of "hothouse children and the saturation of resources that produce them."[37] Often intersecting with "concerted cultivation," "hothousing" refers to attempts to both accelerate and intensify a child's early knowledge development: education becomes a means of investing in the child's future, causing and enabling them to compete at younger and younger ages to, for example, get into schools that will secure their success. Such figurations of the child as "accumulation strategy," to use Katz's phrasing, fortify and are fortified by the specter of the child as waste—that is, the "surplus" or excess child "managed" through criminalization, super-exploitation, and organized violence (see chapter 4).[38]

Putting aside the important metaphysical and ethical questions that should prevent the straightforward conflation of the fetus with the figure of the child, apropos of the temporal crisis of childhood identified by Rekret, it is significant that prenatal sound systems and the discourses associated with them erode any clear distinction between prenatal and postnatal development. Indeed, if, as Rekret suggests, capitalism's entry into the home has reconfigured temporal categories of child and adult in the process, prenatal sound systems evidence how capitalism also enters the reproductive body and the "first home" of the uterus.[39] Affirming Katz's diagnosis of the child-as-accumulation-strategy, prenatal sound systems deploy auditory materials to enable investment in the future-child prior to birth: they serve to shape the future-child and with them the future of the child. Furthermore, in alignment with Heckman's proposition that the earlier the investment, the greater the return, prenatal sound systems seek to develop human capital in utero and, in so doing, accelerate childhood development. In this regard, "returns" are attached not only to what a child can do but when they can do it. The listicle "9 Best Prenatal Sounds and Systems for Baby and Mom" by the website The Pregnancy Zone is exemplary in this regard. Under the subheading "because your baby deserves the best chance of success," the article claims that prenatal sound systems "give your baby the advantage he or she needs by helping them learn even before they're born."[40] The advantage referred to in this description is instructive: What, precisely, are unborn infants gaining an advantage over? The implication, it would seem, is that it is never too early to prepare for a competitive future.

## IMPRESSIBILITY AND THE LISTENING FETUS

Fetal life is frequently mediated by sound technologies. Sonography translates the echoes produced by high frequency (inaudible) sound waves into images of the fetus. Doppler devices are used on obstetric wards and by community midwives to listen to the fetus's heartbeat and monitor their well-being. Like sonography, these devices use ultrasound waves to reflect off moving blood vessels and simulate a sound. These technologies have become increasingly available outside of medical settings. Private providers offer 3D and 4D "bonding" ultrasound scans, allowing pregnant people to purchase photographs, videos, and other keepsakes featuring images of their unborn child. There is also a significant commercial market of dopplers and other prenatal amplification devices that allow prospective parents to listen to and, in some instances, record the fetal heartbeat at home. These devices are advertised as having various features and benefits. A fetal doppler by Sneakpeek (a company that sells gender prediction kits) is described as allowing a pregnancy to become "even more real" by allowing prospective parents to listen to their unborn child's heartbeat either through a speaker or, for a more "intimate" experience, through headphones.[41] BabyDoppler's Sonoline B device, which can be purchased in pink, blue, or teal, offers prospective parents a "enhanced pregnancy journey": it features what the company dubs "intelligent noise reduction" and comes with an app that allows recordings to be shared with others.[42]

The auditory construct of the fetal heartbeat is subject to political contestation. In the US, restrictive legislation aiming to implement abortion bans from six weeks have been referred to by proponents as "heartbeat bills." Organizations such as the American College of Obstetricians and Gynaecologists have made clear that, from a medical perspective, there is neither a fetus nor a heartbeat at six weeks gestation: at this stage, pregnancy concerns a conceptus or embryo, whose "cardiac activity" consists of a group of cells with an electrical signal. As Rebecca Lentjes makes clear, "heartbeat bills" rely on the animating capacity of sound to enliven embryos and fetuses: the animacy ascribed to the fetal heartbeat through legislation relies on the production of the pregnant person as less animate. The political amplification of the fetal heartbeat, in other words, silences the pregnant person.[43]

A similar process of amplification and silencing occurs with commercial fetal heartbeat dopplers. The fetus is imagined as having a discrete auditory existence that can be listened to with ease: these devices are presented as a means of listening to "your baby" instead to one's own body. Not only does this silencing of the pregnant person's body conform to politically regressive conceptualizations of pregnancy, it also creates unrealistic—and risky—expectations of how easy it is to identify fetal sounds. There are now several campaigns that seek to raise awareness of the dangers of home dopplers, given that it is difficult to distinguish the simulated sounds of the fetus's cardiac activity from those of the pregnant person's body and the uterine environment. As a result, the devices can offer false reassurance of fetal well-being if the pulse of the pregnant person or placental blood flow is mistaken for the fetal heartbeat. These campaigns have proposed that dopplers should be used in combination with the trained, expert listening of healthcare professions, who can accurately and reliably discern the fetal heartbeat from other uterine sounds.[44]

Dopplers present the fetus as that to which one can listen: through a process of sound simulation their presence—and wellbeing—is made audible. The auditory construct that arises from this process serves to produce the fetus as an aurally and ontologically distinct individual. Prenatal sound systems, by comparison, establish the fetus as a distinct individual through their capacity to listen. This fetal listener's reactions to sound stimuli are not simply autonomic: Babypod suggest that fetal reactions to their device are "responses" rather than driven by reflex, while BabyPlus describe "prenatal children" as being able to compare and discriminate between sounds. Taken together, these accounts present the fetal listener as an active and engaged recipient of auditory stimuli. However, both Babypod and BabyPlus suggest that this "active" listening is reliant on either the fetus's ability to hear music clearly using the method of vaginal mediation (in the case of the former) or their exposure to a specifically designed, proprietary curriculum (in the case of the latter). In other words, the fetus's ability to become an active and engaged listener relies on specific interventions into the uterine environment.

The association drawn by prenatal sound systems between fetal listening and neurological and cognitive improvement not only invokes much popular thinking about music's benefits. It also has resonances with eugenicist ideas and practices: particularly, those described as the "new" eugenics, which have often found articulation in assisted reproduction. For its advocates, new or liberal eugenics significantly departs from the "old" authoritarian eugenics of the early twentieth century. Where old eugenics were reliant upon poor science, racism, and state coercion, more recent possibilities of human enhancement through genetic and other reproductive interventions are voluntary and based on the individual choices, preferences, and values of parents-as-consumers.[45] The freedom to make genetic selections is considered akin to private, parental choices regarding education and nutrition: if the child is recognized as the property of their parents, then the right to make decisions relating to their genetic makeup coheres with the right to procreate and the right to direct a child's upbringing.[46] Yet attempts to rescue a new liberal eugenics from the old fail to account for the persistence of class-based, disability-based, and racialized hierarchies of human value, traceable in both the makeup of donor markets and the legislative and economic barriers to assisted reproduction.[47]

Although it is infrequently acknowledged, auditory and eugenicist histories have intersected. The term eugenics was coined by the English polymath Francis Galton as an adaption of the Greek words for "well born."

In his 1869 study *Hereditary Genius: An Inquiry into Its Laws and Consequences*, Galton studied musicians to bolster his claim that exceptional abilities tend to run in families, citing the composers Bach, Haydn, Mendelssohn, and Mozart as instances of hereditary genius. Following this, Galton turned his attention to the variability of human traits. Galton proposed that sensory discrimination was correlated with intelligence, with greater capacity for sensitivity exhibited among the most intelligent. As part of this work, Galton developed methods of anthropometric testing—measurements quantifying features that Galton understood as markers of ability and behavior. Aural discrimination features in these batteries, with Galton testing keenness of hearing and pitch perception.[48]

In the early twentieth century, the notion of musical talent as hereditary informed the establishment of "musical eugenics" in the US. The music psychologist Carl E. Seashore, whose "Measures of Musical Talent" became influential in education settings, worked closely with the American Eugenics Society. From 1921, in collaboration with the Eugenics Record Office under Charles Davenport, Seashore undertook a decade-long experiment at the Eastman School of Music that sought to validate his measurements of musical aptitude and interrogate the ways in which insights from eugenics research might be applied to musical education. The Eastman School adopted Seashore's measures—a battery of listening tests requiring the listener to distinguish increasingly similar musical sounds—as part of their admissions process, and to determine those more "capable" of receiving musical training.[49] As Alex Cowan argues, Seashore's Measures of Musical Talent are part of a broader history whereby standardized intelligence testing is used as instrument of racial and class-based social stratification: they were deployed in explicitly racialized studies of musical ability and were used to lend support to the extension of educational segregation.[50]

Eugenics also features in the history of sound's mechanical reproducibility. The telephone's inventor, Alexander Graham Bell, was involved in the US eugenics movement at the beginning of the twentieth century and played an important role in emerging biopolitical approaches to deafness. An advocate of oralism and teacher of deaf children, Bell sought out technological solutions to deaf communication. His ear phonautograph, which he developed with the ear specialist Clarence J. Blake, was intended to transform the vocal sounds of deaf speakers into visible tracings, so as to aid their adoption of "correct" speaking sounds. Bell was working at a time when the deaf community had become increasingly established, with residential schools and sign language facilitating regular interactions between deaf people. At the same time, deaf people increasingly embodied cultural anxieties about "foreignness" in the US: in the late nineteenth century, the deaf community faced stigmatization as one of a number of "insular" and "alien-appearing" populations that were deemed harmful to the well-being of the nation and to its own members.[51] While Bell and Blake's ear phonautograph was ultimately unsuccessful as a pedagogic tool, its underpinning acoustic research contributed to the development of the telephone. Thus, as Jonathan Sterne argues, sound reproduction arose, in part, from attempts amongst hearing people to rectify the cultural "problem" of deafness by training the deaf to pass as hearing people through speech.[52] Put differently, the origins of sound reproduction are connected to a desire to avert the reproduction of deafness through cultural and communicative assimilation. Indeed, after conducting genealogies of deaf families and surveying deafness in schools, Bell advocated for the prevention of deafness through, among other strategies, discouraging marriage between those born deaf.[53]

These audist and ableist histories find new resonances in some of the suggested benefits of Babypod and BabyPlus. A study by Institut Marqués concludes that the Babypod speaker could potentially be used for conducting prenatal hearing screening, although a rationale for this screening is not offered. The company has also claimed that Babypod can be used to eliminate dyslexia and other forms of cognitive difference. Brett Logan, the inventor of BabyPlus, has similarly speculated in his self-published book, *Learning Before Birth: Every Child Deserves Giftedness*, that his prenatal learning system could be used to mitigate autism, certain types of schizophrenia, and Down syndrome.[54] The accuracy of such claims is not the point: they are significant in that they illustrate an aspiration for sound, music, and their mediating technologies to work against cognitive and neurological difference.

While their positioning of disability and neurological difference as requiring mitigation in the pursuit of intellectual enhancement has some overlap with eugenicist ideas, prenatal sound systems otherwise appear to have little to do with the eugenics of Galton, Davenport, and Bell or indeed with the new eugenics associated with selective sperm banks and egg donor markets, both of which concern the minimization and maximization of hereditary traits via selective reproduction. By contrast, devices such as Babypod and BabyPlus suggest that the enhancement of the fetal listener is primarily achieved through environmental influence. The BabyPlus website is instructive in this regard, claiming in the "About" section that "a recent study published by UCLA researchers states only 45% of the temporal lobe, the part of the brain which plays a role in learning and memory appears to be inherited. In other words you can make a difference in your child's foundation for learning. That's empowering!"[55]

In siding with environmental impact over genetic destiny, neurological plasticity over biological determinism, prenatal sound systems can be more accurately described as enacting a biopolitics of impressibility. Writing in the context of the nineteenth-century US, Kyla Schuller identifies the emergence and circulation of a pre-geneticist discourse of impressibility among scientists, writers, and reformers. Pertaining to the capacity for bodily transformation in relation to external agents, impressibility came to differentiate the "vitality" and vulnerability of the civilized—that is, those "unmarked" as white and other racial groups deemed capable of "progress"—from coarse, ridged, and prehistorical elements of the population. Through the management of the body's environment, experience, and its impressions—that is, affective relations that leave a trace—certain populations could be improved, depending on their capacity for impressibility. Youthfulness was deemed a stage of heightened impressibility, and as a consequence, the nervous system of uncivilized youth could be impressed with beneficial sensations, seeking to morph their heritable material to the benefit of the nation. Early twentieth century eugenicists positioned themselves in opposition to those who believed that hereditary materials could be changed experientially, however historical and ideological coherences remain between the two: in seeking to regulate, modify, and improve hereditary materials at both the level of the individual and the collective, advocates of impressibility might be best understood as an antecedent to eugenicists' goals.[56]

Ideas of impressibility also informed cultural and medical perspectives on pregnancy and the maternal-fetal relationship, whereby the emotions, experiences, and behaviors during pregnancy were understood to influence a child's development and were used as an explanation for disability. These discourses held influence on both sides of the Atlantic, however, they persisted for longer in the US: even as such theories were discounted in the UK and Europe, and science and industrialization introduced new concepts for what occurred inside bodies—the fetus, for example, became a named entity in 1870—many US doctors clung to older ideas about the maternal influence on fetal development. The notion of maternal impressibility served to maintain the anxiety of pregnant women, particularly white women, making them responsible for ensuring their child's "normalcy."[57] A 1905 article entitled "The Maternal Impression Superstition, and Why It Survives" by Kansas doctor Edwin Taylor Shelly makes apparent such associations, arguing that physicians should consider the doctrine of maternal impressions a stigma on the profession since as it lacks scientific merit, and furthermore, causes fear in the expectant mother:

> Of course, when impressionism leads the expectant mother, only to indulge, to a greater extent than usual, her taste in music, books or works of art, the belief is nothing more than a harmless, introspective, fad; but because of the fact that, too often, her days and nights are filled with thoughts of marks and monsters, the time has surely come to enter fervent protest at this crass belief—this hideous, lingering apparition of a starless night when Science slept.[58]

Shelley's concluding remarks make clear that music—alongside other forms of cultural practice and consumption—were adopted by those seeking to positively exert influence on their unborn child. This association is similarly traceable in other texts from the era. For example, in Mary Ries Melendy's 1904 *Science of Eugenics*, a manual on marriage, motherhood, and childrearing, expectant mothers are told to "carefully select their company, their books, their thoughts and their ideals" and attend to their reading and study: "Remember that the mother's superior culture in music, art, or any study, must be exercised during gestation, if it is to have any effect on the child."[59] Claiming that "it will be found that persistent mental force . . . is a stronger power than heredity," Melendy details an instance where a prospective mother who lacked musical ability was determined that her child should be better endowed than her in this respect.

The mother took music lessons during the final months of pregnancy. As a result, the child learned music easily. Melendy thus advises that "whatever is lacking in yourself which you would see in your child, practice assiduously, and the longer before conception this is begun, the better."[60]

Ideas of maternal impressibility continue to hold significance in the twentieth and twenty-first centuries, although their expression has been transformed by changing scientific, technological, and economic milieux. As Sarah Richardson has shown, by the 1920s and '30s, the nineteenth century concept of maternal impressions had been largely rejected by the medical establishment, with prospective mothers being assured they could not alter their child's traits. However, anomalies in patterns of inheritance in data about plant and animal genetics eventually gave rise to the doctrine of "maternal effects": an ambiguous term first coined in 1935 in a paper on fruit fly genetics, which came to refer to a range of nongenetic, maternal influences on future generations.[61] Holding explanatory power for various birth disparities and informing public health policy and guidance, the doctrine of maternal effects placed responsibility for the health and well-being of future generations on the shoulders of the pregnant individual.[62]

The doctrine of maternal effects has found renewed expression with the rise of epigenetics: an emerging, postgenomic field, which, at its broadest, concerns the interaction between genes and environment and the relationship between environmental factors and genetic expression. Its rise has been cautiously welcomed among some feminist and decolonial scholars, insofar as it offers a means of articulating how structural social harms—such as the violent dispossession endemic to coloniality, or inequitable exposure to toxicity due to racial inequalities—find expression on a biological register. Others have used epigenetics to complicate propertied, heteronormative notions of inheritance that underline nuclear kinship arrangements. Addressing the gestational labor of commercial surrogates in the context of queer family making, Jaya Keaney has employed epigenetics to challenge the assumption that genetic transmission exclusively constitutes racial inheritance: if the environmental effects of race and racism leave a genetic trace, then the separation ordinarily assumed by commercial surrogacy arrangements—between racializing gametes and nonracializing womb—is undermined.[63] Yet despite these feminist interrogations, in practice, the application of epigenetics has often reinforced maternal responsibility and blame.[64] While what "counts" as environment for epigenetics is rather capacious, as Natali Valdez illustrates in her ethnographic study of how epigenetics is translated into clinical prenatal interventions, the uterine environment has received disproportionate attention: to this end, pregnant people's bodies are differentiated as "risky" and "optimal" environments, generated by individual behaviors and personal choices, to which the health outcomes of future generations are attached. Translated in a "scientific and economic milieu that prioritizes individual responsibility and risk" and "collapses the environment into discrete variables," epigenetics results in an intensification of maternal responsibility, over and above other forms of environmental damage or nurture, leaving unrealized a reimagination of a maternal environment that "(re)distributes responsibility across bodies, communities, and government."[65]

While it would be erroneous to suggest that in seeking to enact cognitive improvement, prenatal sound systems have much in common with an epigenetic model of environment-genome interaction, they nonetheless share with some clinical translations of epigenetics a recognition of the uterine environment as a key site of intervention that can be improved to the benefit of the fetus through "appropriate" personal choices. More generally, these technologies highlight the continued salience of ideas of maternal impressibility, whereby the prenatal environment has lasting postnatal effects. In understanding the fetal listener as that which can be cognitively transformed via its auditory affections and in offering the pregnant person the opportunity to "optimize life" through the provision, management, and control of sound sensations in utero, these reproductive sound technologies mark one afterlife of impressibility, to augment Schuller's phrasing. Indeed, they testify to the coherence of neoliberal figurations of human capital and biopolitical figurations of impressibility, insofar as both figure subjectification in terms of malleability, responsiveness, and enhancement. However, where nineteenth-century notions of impressibility were bound up with questions of heredity and the improvement of the race, the relation between impressibility and heredity is reconfigured by prenatal sound systems and their fetal listener: they can be understood to partially replace biological with financialized notions of inheritance, where competition, success, and optimization pertain to familial and intergenerational investment.

The "egalitarian" implications of a shift away from biological "destiny," are highlighted by BabyPlus's Brett Logan: in *Learning for Birth,* Logan he suggests that the efficacy of prenatal enrichment means that any baby can—in theory—benefit from BabyPlus. Yet at the same time, he draws on notions of genetic and biological inheritance to conjure a "competitive future" attached to the rising threat of overpopulation. Invoking, once again, the child-as-accumulation strategy, Logan posits that responsible and caring parents would elect to enrich the fetus so as to "insure for their genetic successors the fullest outcome."[66] It is hereditary bonds, and the desire for those genetically linked to us to succeed, that are presented as motivations for prenatal enrichment. Furthermore, Logan proposes that BabyPlus could contribute to measures against the threat of overpopulation by encouraging families—particularly those in the "developing" world—to choose quality rather than quantity when making reproductive choices: "Women who give birth to pre-natally primed offspring can see what their additional resourcefulness means to the family dynamic, whether greater child wellbeing, enhanced interpersonal rewards, or higher income.... These assets in one individual can weigh against bearing many babies of lesser potential."[67] Meanwhile, those who fail to choose prenatal enrichment should be aware of the "superlative competition fostered by the complex and crowded world already threatening even projected limits."[68] Unsurprisingly, the neo-Malthusian specter of fierce human competition in an overpopulated world—a problem for which "developing" nations hold disproportionate responsibility—does not appear in the commercial marketing materials for BabyPlus.[69]

The diminished emphasis on genetic heredity that may accompany financialized notions of inheritance, then, should not be mistaken for a postracialism, nor indeed does it mark a departure from its ableist articulations. Indeed, in keeping with the dominant optics of assisted reproduction technologies, whose "magic and promise" has often been represented by the image of the healthy white infant,[70] white women and smiling white babies feature heavily in the marketing materials of Babypod and BabyPlus. Yet the racialization of these devices extends beyond their visual and indeed sonic signifiers (we might recall here the rejection of "African drums" and the preference for Mozart, described by Babypod). If impressibility is enmeshed with a history of racial differentiation, which distinguishes whiteness as the result and marker of the most advanced capacity for plasticity, adaptability, and transformation, then it is unsurprising that the malleable fetal listener attaches itself to a white-racialized aesthetic.

## UTERINE AUDIOPHILIA

In writing about sound and music, gestational metaphors are frequently employed as a means of conjuring auditory experience. Various "immersive" sonic environments are figured as evocative of fetal life within the womb, and a return to presubjective envelopment. Popular music writer Simon Reynolds, for example, repeatedly makes reference to uterine space when referring to atmospheric or ambient sonic encounters. Writing about rave, he describes the experience as "a nowhere/nowhen wonderland, where time is abolished, where the self evanesces through merging with an anonymous multitude and drowning in a bliss-blitz of light and noise. It's a regressive womb-space or clandestine kindergarten." Dub, meanwhile, is described by Reynolds as invoking "the blurry sonic intimacy of womb-time, the lost paradise before individuation and anxiety."[71] Other depictions of the sonic as uterine are present in the work of philosophers such as Jean-Luc Nancy and Peter Sloterdijk, as well as the Canadian acoustic ecologist R. Murray Schafer.[72] In his chapter on the natural soundscape, Schafer presents an origin myth of listening and what he terms the "first sound": the caress of waters. Where Greek myth tells of how humans arose from the sea, Schafer notes how

> the ocean of our ancestors is reproduced in the watery womb of our mother and chemically related to it. Ocean and Mother. In the dark liquid of ocean the relentless mass of water pushed past the first sonar ear. As the ear of the fetus turns in its amniotic fluid, it too is tuned to the lap and gurgle of water.[73]

The sea is thus "a fertile sonic archetype. All roads lead back to water."[74] For Schafer, the sonic ocean-womb is evoked by listening experiences in which distance and directionality are absent—a condition he hears facilitated in much contemporary and popular music:

> The listener finds himself [*sic*] at the center of the sound; he is massaged by it, flooded by it. . . . where then is the dark and fluid space from which such listening experiences spring? It is the ocean-womb of our first ancestors: the exaggerated echo and feedback effects of modern electronic and popular music re-create for us those echoing vaults, the dark depths of ocean.[75]

As Schafer's remarks attest, the auditory construct of the womb deployed as evidence of the primacy of aurality. Hence cultural theorist Dominic Pettman opens his account of sonic intimacies in a manner akin to Schafer: "We are born *in* and *of* sound. Our first prenatal experience is overwhelmingly aural: we become embodied and enfleshed within the squelches, rumbles, and pulsing thumps of the mother's body."[76] Film sound designer Walter Murch similarly posits aurality as the first sense, proposing that the disembodied voice carries "an echo of our earliest experience of the world": it reminds us of the revelation at birth that the "song" from the beginning of the listener's consciousness in the womb is in fact "the voice of another and that she is now separate from us and we from her." This sound is a reminder of a loss: "We regret the loss of former unity—some say that our lives are a ceaseless quest to retrieve it—and yet we delight in seeing the face of our mother: the one is the price to be paid for the other."[77]

This formative connection between sound, aurality, and the earliest stages of life has been subject to commodification. In the 1970s, Japanese obstetrician Hajime Murooka released a series of recordings of uterine sounds intended to help soothe and calm babies. The LP *Lullaby from the Womb* (titled in the UK as *Sleep Gently in the Womb*) was released in 1974 by Toshiba EMI. The A side features a series of uterine sound recordings—track titles include "Sound from Aorta (A)" and "Blood Vessel of Navel Chord and Placentas"—some of which are overlayed with music by Schumann and Wolf-Ferrari. Side B is titled "Familiar Music for the Baby's Environment": intended for older infants, it consists of extracts of "relaxing" instrumental music, including Bach's "Air on the G String" and Tchaikovsky's "Panorama" from the ballet *The Sleeping Beauty*. Murooka's release preempts the array of baby sleep apps, playlists and recordings now available to prospective and new parents, many of which offer "womb sounds" to calm, soothe, and quiet babies (see chapter 5).

More recently, Sonic Womb Productions, "a multidisciplinary collaboration between scientists, engineers and sound artists" has sought to recreate the experience of hearing in utero, using an "acoustically optimised enclosed space." A collaboration beginning in 2012 between fetal medicine researcher Eric Jauniaux, sound studies scholar and artist Julian Henriques, and venture developer Aude Thibaut, the project has developed a "fetal filter" sound system. This is installed within the Orbb: a well-being capsule intended to help users relax, regain focus, and improve performance.[78] There are multiple uses suggested for the Sonic Womb sound system, including the research and creation of uterine soundtracks for therapeutic, educational, and relaxation purposes and the development of the "next generation" of neonatal incubators for premature infants. Affirming, once again, the attribution of maternal responsibility and risk to the individual, it is also proposed that the technology and its associated research could help pregnant people make "choices" about the sonic environments they "expose" the fetus to: "The lack of understanding of what a fetus hears in utero . . . hinders mothers-to-be from making fully informed decisions about modern artificial sounds to which her unborn child can be safely exposed."[79] Yet beyond such uses, the Sonic Womb is understood to have a wider appeal, in enabling a listener to reexperience a time in their lives they cannot remember. After all: "who wouldn't want to return to the origins of perception in the fetal state where hearing . . . is the only sense connecting consciousness to the outside world, a place long forgotten but eagerly rediscovered . . ."[80] Sonic Womb thus frames uterine sonic experience as simultaneously a product of individual maternal choices and a universal, transcultural beginning, insofar as attempting to recreate the spatial and timbral attributes of the fetal sound environment takes listeners back to "the origins of perception": the time of aural primacy.

These various aural and written depictions of sonic experience in the womb exemplify what I want to call *uterine audiophilia*. Whereas the audiophile wants the recording to reproduce the "lost paradise" of live sound, for Reynolds, Schafer, and others, sound reproduces the lost paradise of the womb: the comforting, immersive space of belonging. However, just as the audiophile values the reproduction of an *imagined* original sound, uterine audiophilia concerns the reproduction of an *imagined* prenatal experience, one which is reliant upon omitting the substantive differences in prenatal and postnatal perception. The references to "consciousness" and "listening" in utero, for example, belie that the fetus is mainly asleep and unconscious, and at a stage of cognitive development that cannot be recreated ex utero.[81] Put differently: while it is possible to approximate the bodily, vocal, and environmental sounds as they are mediated by and through the uterus, it is not possible to "hear" as a fetus does. Furthermore, in approaching the pregnant person's body as material container, resonant chamber, or sonic envelope and in prioritizing "content" and its perception (i.e., fetal listener and uterine soundscape) over "medium" (pregnant person), uterine audiophilia's conflation of sonic experience with uterine experience is indebted to the long-standing reduction of pregnant people to their capacity for maternal provision.

Feminist critics have long challenged these objectifying accounts of pregnancy and gestational labor as they appear in relation to reproductive technologies such as ultrasound. Carol Stabile, for example, has traced the figuration of the maternal body as both benevolent environment and inhospitable wasteland. Writing about the development of fetal photography and the correlative ideological work performed in the service of US new right politics in the 1980s and 1990s, Stabile notes how emergent representations of "foetal personhood" rely upon the erasure of the pregnant person, reducing them to "passive, reproductive machines."[82] Whereas historically, appeals to motherhood worked to erase female subjectivity and sexuality, the unprecedented establishment of "foetal autonomy" via visualizing technologies enables the pregnant person's body to become "an environment that the foetus alone occupies."[83]

A similar figuration is produced by uterine audiophilia, whereby the pregnant person appears as a benevolent (sonic) environment for a fetal "listener." As a result, uterine audiophilia is often attached to a romanization of the fetal–uterine connection: as the comments by Reynolds, Schafer, and Murch attest to, the relationship between fetal listener and their listening environment is one of unity. Yet what this obscures is the complex and often antagonistic dynamics required by pregnancy. Challenging characterizations of pregnancy as involving two discrete selves and/or cooperative occupation, Sophie Lewis has emphasized the multidirectional but nonetheless asymmetric character of human gestation, whereby fetus and gestator are both transformed in the process. Although the fetus is part of its gestator, this is by no means a straightforwardly happy union:

fetuses distinguish themselves and they "do so brutally." Humans can, and often do, die of gestation. Take Schafer's sonic archetype of water. While the originary water might be fine for fetuses—their lungs are filled with it, they neither breathe nor drown—as Lewis makes clear, for the person filled with that water, it can be extremely dangerous:

> A person does not suddenly become an amphibian by virtue of becoming pregnant. Yet she (usually a she) is flooded from the inside: control of the circulation overridden, arteries jammed wide-open, blood pressure forced into overdrive. A plug forms, to seal as much as a liter in the vessel that is the amnion, the placental tank.[84]

Fetuses and gestators neither exist autonomously from one another nor do they exist in harmonious oneness. This "less than two, more than one" relation of human gestation is aggressive and risky. And yet, inasmuch as the interior cannot be cleanly separated from the exterior, biological from social, uterine space from wider environment, the degree of risk that human gestation carries with it is coconstituted by other factors. Gestation is far less risky when undertaken in the context of universal healthcare and robust reproductive rights than when these are lacking, for example. It is made far more risky by medical racism, profit-oriented healthcare, and poor-quality post-partum care, all of which are contributing factors to what has been labeled the US's "maternal mortality crisis."[85] Gestation is thus situated within a broader matrix of constitutive relations that incorporate but extend beyond fetus in utero and the personal "choices" of the pregnant person.

Uterine audiophilia disavows the social production of pregnancy by figuring the prenatal sound environment as ahistorical, natural, and universal: it is an originary "echoing vault" that the fetal listener alone inhabits. Indeed, uterine audiophilia coheres with what Annie Goh has described as sonic naturalism, referring to sound studies' fixation on both the nature and naturality of sound: we might recall here Sonic Womb's suggestion that its high-tech and highly mediated reconstruction of the uterine soundscape may offer mothers-to-be insights into the risks of fetal exposure to "modern artificial sounds." A cisheteropatriarchal construct, sonic naturalism is underpinned by Eurocentric dualisms that distinguish masculinity and femininity, object and subject, nature and culture.[86] In its framing of womb as a presocial sonic environment and aurality as the fetal listener's "first sense," uterine audiophilia affirms the naturality of both sound and pregnancy.

In figuring the pregnant person as a sonic environment for a listening fetus, uterine audiophilia finds expression in the design and circulation of prenatal sound systems. However, its specific iteration differs between devices. In the case of Babyplus and Babypod, these differences are traceable in both their approach to sound and mediation. As noted earlier in this chapter, BabyPlus uses sounds similar to those produced by the maternal body during gestation to create contrasting in utero and ex utero rhythms. According to Logan, BabyPlus marks a continuation of long-standing maternal practices involving sound:

> Prenatal stimulation merely recasts in contemporary format venerable music revering a new member of the human tribe; before the healthcare community started regarding pregnancy as a problematic condition, gravid mothers danced to chants or drumbeats indelibly touching their unborn babies—which is the reason postnatal infants fix an avid ear near a parent's heart... By simply supplementing through technology the essential influence—a direct descendant of a luckily redundant and therefore imprintible cardiation—an ancient clue to multifactorial capacity has been furthered with highly efficacious lifespan repercussions.[87]

Combining New Age romanticism, scientistic terminology, and techno-optimism, Logan's account of prenatal learning is a narrative predicated on both notions of sonic surrogacy and sonic naturalism. It fetishizes the ancient and Eastern, noting that Western history has had little concern for gestational matters, while also recognizing technology as mechanizing and enhancing "natural" auditory practices associated with reproduction.[88] Sound reproduction technologies are presented as a means of strengthening the ancient and essential influence of sound *in utero*: Logan suggests that Babyplus is akin to the (then) common use of the Sony Walkman by expectant parents to play music to the fetus by placing headphones on the pregnant person's abdomen, the invention of which is listed on the BabyPlus website as a key milestone in the history of prenatal learning.[89]

Where BabyPlus seeks to "supplement" the sound environment of the fetal listener, Babypod aims to overcome it. If the fetal "autonomy" afforded by visualizing technologies is predicated upon the disappearance of the pregnant person, Babypod provides an auditory analogue to this process of disappearance inasmuch as the device attempts to make the necessary presence of the pregnant person inaudible. Babypod claim that their vaginal speakers differ from other prenatal sound systems since they enable the fetus to have a higher fidelity and thus a more effective learning experience. The company claims: "We have discovered how to allow your baby to hear like us, in such a way that they will perceive sound effectively in terms of intensity, and free from distortions." Hearing music clearly and "like us," according to Babypod, is "solely possible via the vagina" because "the abdominal wall muffles sounds." The website states:

> The uterus is a place protected from the exterior, and it is the mother's body that carries out this protecting role through multiple layers of soft tissue. These attenuate the intensity of sound and distort it in its journey to the uterus; it's similar to what happens when you hear a conversation in the next-door room without catching everything that is said.... By placing a speaker inside the vagina, we overcome the barrier formed by the abdominal wall and the baby can hear sounds with almost as much intensity and clarity as when emitted.[90]

In other words, Babypod considers itself superior to its competitors, since it minimizes the noisy presence of the pregnant person's body. It thus coalesces with audiophilic notions of sonic realism that are predicated upon the dream of mediumless—and therefore noiseless—sound.

Fidelity discourses tend to imagine recorded sound as a copy or documentation of a live event. Such discourses fail to recognize, as Alexander Weheliye summarizes, that "no recorded performances, not even live recordings, are 'real'—or even representations thereof. Rather, they are virtual productions created through interactions of musicians and listeners with recording and reproduction technologies."[91] Yet for the audiophile, the "original" event (inevitably a sonic fiction) is a crucial index. As Aden Evens notes, for the audiophile, "the sole criterion for judging recorded music is the extent to which it sounds like live music." Consequently, "the audiophile trains his [*sic*] ears to sort out the 'true' sound of the violin from the artifacts introduced by the recording and playback process. Audiophiles are the self-appointed guardians of fidelity."[92] High-fidelity recordings and playback are those in which the medium's noisy presence remains unnoticeable to the listener; whereas low fidelity refers to the way in which the medium draws attention to itself. As Evens highlights, audiophile discourse is masculinist: its history is indebted to the musical activities, practices, and perspectives of predominantly white, male, and wealthy listeners. In masculinist audiophile discourse, furthermore, the containing and sustaining medium is feminized: it is frequently imagined as passive, receptive, empty, and awaiting inscription and control.[93] Yet the medium will inevitably leave its noisy impression on sonic content. For the audiophile, the best medium would be no medium. As this is a material impossibility, the "best" sound technologies are those that reproduce the (imagined) live event with little discernible change; they pay fidelity to the (imagined) original sound. In auditory cultures organized around audiophilic notions of fidelity, then, the sonic medium—the means of reproduction—should work to provide an illusion of its own absence, in spite of its necessity. Babypod, likewise, minimizes the audibility and affectivity of the containing technology that is the pregnant person's body, and in doing so exemplifies the paradoxical predicament of the sonic medium: to be there but sound not there, to be essential yet sound absent.

One of the ways that feminists have responded to the ontological objectification, erasure, and devaluation of pregnant people is to denaturalize pregnancy as a component of what Lewis terms "gestational labor," foregrounding its imbrication with a wider matrix of reproductive relations and troubling its conflation with motherhood. As Lewis makes clear, not all forms of gestation end in motherhood, not all mothers undertake gestation, and not all gestators are recognized as mothers. To this end, gestational labor necessarily includes miscarriage, abortion, and surrogacy within its frame.[94] Stabile similarly concludes from her examination of ultrasound technologies that both pregnancy and mothering need to be recognized as nonidentical but interconnected forms of work:

> Although feminists must insist that pregnancy is not identical with mothering, they must also insist that both are "biosocial" experiences—that pregnancy, like mothering, is something that occurs within a specific social, economic, cultural and historical environment and that the experience of pregnancy, as such, is structured by social relations. It is work that women may not, or may, decide to undertake.[95]

Leaving aside this apparent framing of pregnancy as something women choose to undertake—the 2020s backlash against reproductive rights within and beyond the US would suggest that this work is often coerced—the recognition of pregnancy as work is not intended as a means of transferring "respect" or "nobility" sometimes associated with the wage to the exhausting exertions of gestation. In any case, this "respect" is frequently denied to those gestators who most clearly undertake pregnancy as work: commercial surrogates are subject to misogynistic stigma, contributing to the diminishment of their working conditions.[96] Rather, the framing of pregnancy as work/labor is intended to undermine the pernicious separation of uterine relations from the broader matrix of biochemical, material, political, and historical relations that lie behind this lost sonic paradise, to augment Reynolds's descriptor: the bodily exertions, routines, practices, infrastructures, and relations of containment, maintenance, and facilitation through which a life is *made*.

It is striking that the prenatal sound systems contain a partial and necessarily compromised acknowledgment of pregnancy-as-gestational labor. Consequently, these technologies underline that the feminist rearticulation of pregnancy in such terms is only useful as a starting point (and to be clear, both Lewis and Stabile treat it as such): as a grounds for radically reimagining how a life might be safely, collectively, and justly made and cared for within, against, and beyond the demands of capitalism. On the one hand, prenatal sound systems tend to repeat conservative imaginations of the pregnant body as passive container or environment, positing the uterus as a space to be filled. with sound or erased as a noisy barrier to the fetal listening experience. Yet on the other hand, in offering prospective mothers an auditory tool to enact and assist in their reproductive work—a means of shaping, enhancing, and investing in the necessarily competitive fetal listener-as-human capital—prenatal sound systems bring forth a figuration of gestation as the product of technologically aided maternal labors. There are, therefore, important differences between the "sonic naturalist" uterine audiophilia of sound studies, and the uterine audiophilia of prenatal sound systems. Although both foreground pregnancy's association with auditory experience, the gestational metaphors of the former posit sound as facilitating a nostalgic and romantic *return* to an originary listening space. Prenatal speakers, conversely, are future-facing, attempting to accelerate the development of the preborn future-child. Likewise, while both erase the pregnant person and assert fetal autonomy via their capacity to act as a listener, their figurations of uterine spaces differ. Where sound studies' gestational metaphors typically present uterine space as an ahistorical benevolent environment, prenatal sound systems present uterine space as a "perfect classroom": a reproductive workplace.

## CONCLUSION: QUEERING THE UTERINE CONCERT HALL

During Dayna McLeod's *Intimate Karaoke: Live at Uterine Concert Hall*, an "interactive, durational performance" with sardonic motivations, the artist's body is transformed into a music venue in order to "queer the uterus as a viable physical space."[97] McLeod repurposes the Babypod for a middle-aged, white, queer, cisgendered, and nonreproductive body (in the sense that McLeod never has had nor will use it to have children), inviting participants to perform their favorite karaoke song in her uterus. The performance takes place across two sites: a large main room, in which a song is performed, and a smaller room, containing McLeod. Karaoke performers can hear a song's backing track, but other listeners within the larger room cannot. The vocal performance and the backing track are transmitted to and "mixed" within McLeod's uterus. External listeners can eavesdrop using a stethoscope and listen to the faint echo of music that can be heard outside her body alongside "special effects" such as intestinal gurgling. McLeod lies down during the performance, positioned in a way that references gynaecological examinations (figure 8).

In these acts of making, mediating, and listening to music, discomfort, vulnerability, and empathy are distributed among performer and audience. Singing for strangers—particularly unaccompanied—makes those performing in the main room vulnerable. McLeod is also made vulnerable in being penetrated by the Babypod speaker, while audience members are made vulnerable in being invited to listen to music through a person's uterus. McLeod makes clear that her ability to use her body to facilitate the "Uterine Concert Hall" is partly reliant on her whiteness and her body being unmarked by disability: she notes that the relative distance

*Figure 8.* Dayna McLeod, *Intimate Karaoke, Live at Uterine Concert Hall*, curated by FADO Performance Art Centre, copresented with SummerWorks, Toronto, 2018. Photo: Henry Chan.

she has from the racist and ableist histories of exhibiting the bodies of non-white and disabled people allows her to undertake the performance in a way specific to her subject position. However, by using her body to stage, transmit, and perform music, and in employing the audience in the performance, McLeod explains, "I was not just emphasizing the physical emptiness of the uterus (i.e., no fetus) but its conceptual and aural capacity to 'give birth' to a kind of queer kinship and art production that isn't heteronormative or reproductive."[98]

Prenatal sound systems' simultaneous invocation of maternal duty (the obligation to use sound to shape, optimize, and invest in the malleable fetal listener in preparation for a competitive future) and passivity (the reduction of the pregnant person's body to a listening environment, and their figuration as both resonant chamber and a source of noise) may seem contradictory. However, rather than simply highlighting the conflicting demands that characterize contemporary motherhood, prenatal sound systems are illustrative, once again, of how the nuclear family marks the convergence between social conservativism and neoliberalism. Specifically, they foreground how neoliberalism's "human capitals" necessitate a gendered division of labor: accordingly, responsibility for the speculative, and in this case, sonic production of the future-child lies with the mother (to-be) and the private household. Their combination of long-established tropes that objectify and erase the pregnant person, reducing them to their capacity for maternal provision, with a (partial) acknowledgement of gestation-as-labor, and their framing of pregnancy as both natural and in need of a technological fix make explicit how conservative ontologies of reproduction are maintained even as the social, economic, and temporal reconfigurations of neoliberalism impact upon the regimens of pregnancy, parenting, and motherhood.

Their intensification of maternal-familial responsibility, combined with the physical invasiveness of these sound technologies is precisely why prenatal sound systems have attracted a performer such as McLeod: they are fertile technologies for a frictional, queer-feminist (mis)use. Despite their explicitly cisheteropatriarchal technopolitics, McLeod's performance implies that prenatal sound systems can be deployed in ways that subvert the uterine audiophilia that informs their design and circulation as commodities, enabling another, decidedly *non*fetal form of listening to come into fruition. To this end, McLeod takes up the idea of uterus as resonant listening environment. However, rather than being a private space for the fetal listener, McLeod's "Uterine Concert Hall" is open to multiple listeners, performers, bodies, and voices. Where prenatal sound systems use sound, music, and their affective capacities to impress upon and improve the fetus, McLeod's performance instead uses Babypod as a tool for gestating collective social bonds through discomfort, intimacy, and vulnerability, forging a particular configuration of the uterine environment as a generative container for "temporary kindship networks."[99] In so doing, McLeod's piece speculates as to what the uterus might be, what it might hold, and what value it might have outside of its cisheteronormative designations and beyond (from the perspective of capital, at least) the renewal of exploitable life.[100] By producing, playing back, and mediating sound, the uterus becomes simultaneously productive, reproductive, and nonreproductive; both empty space and a site of possibility.

# 4 Auditory Revanchism, Antisociality, and Reproductive Respectability

The 1996 New York Police Department *Quality of Life Enforcement Options Reference Guide* provides advice to officers regarding the public's use of radios and other sound devices. If sound from these technologies is audible in front or outside of premises, then police intervention is warranted:

> A sound reproduction device is any sound device or apparatus, or any device or apparatus for the amplification of any sounds from any radio, phonograph, sound-machine or sound producing device, or any device or apparatus for the reproduction or amplification of the human voice or other sounds. Operation of a sound reproductive device without a permit is a violation of Section 10-108 (2) (d) of the Administrative Code, a misdemeanour. An officer may issue a universal summons, returnable to Criminal Court, to a violator of this section and seize the sound reproduction device as evidence.[1]

Published during the tenure of Howard Safir, appointed New York City Police Commissioner by Mayor Rudy Giuliani, the guide provides an insight into policing during this period. The "quality of life" approach referenced in its title emerged with Safir's predecessor, William Bratton, as he sought to restore order to New York's public spaces. The initiative's roots lay in the "zero tolerance" and "broken windows" methods of policing, through which minor infractions and "disorderly" activities are understood to both impede upon the city's liveability and encourage more serious crime. As the extract above makes clear, the unlicenced operation of sound reproduction devices and listening to music in public were deemed infringements according to the "quality of life" program. Other breaches entailed "panhandling" in streets, roadways or subway stations; washing windshields in traffic; disorderly behavior in parks (i.e., camping, washing and leaving personal belongings unattended); public lewdness; open fires; the erection of shelters; and driving cars with excessively loud radios (so called "boom box cars").[2]

The NYPD Reference Guide is emblematic of what the geographer Neil Smith refers to as urban revanchism. Writing in 1996—the same year as the Reference Guide's publication—Smith diagnoses the emergence of a punitive and vengeful turn in US public culture as the pernicious effects of neoliberal economic policies took hold. In his account, 1990s New York under the tenure of Giuliani is the primary exemplar of "the revanchist city," whereby urban social policy and policing is organized against the poor and the marginalized, "cloaked in a populist language of civic morality, family values and neighbourhood security."[3] Whereas liberal urban policy had assumed that governments had some responsibility for maintaining a minimum quality of life for everyone, Giuliani's policies were structured around a vendetta against a constructed urban enemy whose existence was itself a threat to living standards. The adoption of zero tolerance policing, resulting in the criminalization of activities deemed inimical to the quality of life in city neighborhoods, gave the New York City Police Department unprecedented powers to remove unhoused people from the streets. Simultaneously, the administration cut funding for soup kitchens and cleared squatters out of otherwise unoccupied Lower East Side buildings.[4] The homeless were excoriated alongside "workers and 'welfare mothers,' immigrants and gays, people of color, . . .squatters, anyone who demonstrates in public" as people who had stolen the city from the white middle classes.[5]

Smith grounds his diagnosis of urban revanchism by drawing parallels to the revanchist political movement in late nineteenth-century France. A right-wing revenge-oriented politics grounded in anger at the increased liberalism of the Second Republic, defeat in the 1870 Franco-Prussian

War, and the ascendancy of Paris Commune, it sought to take back the country in the name of "traditional values"; the "simple virtues" of honor, family, the army, and the Republic. In New York, Giuliani's revanchism emerges as "self-centred optimism" is replaced by public fear during the 1987 market crash and the long period of economic depression that followed. This crisis unleashed middle-class anger, which, with the help of a media culture propagating fears about urban safety and disorder, came to be directed at marginalized populations.[6] Such affective investments meant that, "rather than indict capitalists for capital flight, landlords for abandoning buildings, or public leaders for a narrow retrenchment to class and race self-interest," Giuliani led a charge against beggars, sex workers, squatters, homeless people, car windscreen cleaners, graffiti artists, and unruly youths as the culprits of urban decline.[7] The vacuum created by the strategic withdrawal of state-managed subsidies and diminishing opportunities for waged work was filled by a market increasingly determinant of "natural" social relations and consequences and an expansion of criminalization and incarceration primarily targeting the city's "surplus" population—that is, an unwaged segment of society simultaneously produced and externalized by capital, whose reproduction is therefore reliant on welfare, participation within the informal economy, or subsistence activities. Some of the surplus population act as the reserve army of labor, what Marx describes as "a mass of human material always ready for exploitation" should capital require it.[8] However, as deindustrialization and automation have decreased the regional and global demand for labor, much of the surplus population finds itself permanently expulsed, existing only to be "managed" by capital and the state, often with recourse to criminalization and incarceration.[9] Revanchism, then, is an expression of the crisis of reproduction as it plays out in urban space, and is imbricated with anxieties about population, contagion, and social dysfunction.[10]

As the NYPD guide attests, revanchism has an auditory component. Sound, music, and their associated technologies have been policed in accordance with revanchist logics. Sound's ability to travel and traverse—to spill out of environments and cross into others—is one reason that it is frequently bound up with contestations over social space. Thus, as Lilan Radovac argues, the dual forces of revanchism and gentrification (i.e., a process of urban "renewal" that often precedes or accompanies revanchism) have meant that marginalized communities are subject to stringent constraints on their aural activities.[11] What I want to name as *auditory* revanchism includes but extends beyond such punitive policing of the sounds of surplus urban life: it also concerns the deployment of sound, music, and aurality as a means of *enacting* revanchist policies. To this end, this chapter addresses revanchist reproductive sound technologies that act as surrogates for police and security forces, coalescing around what Sabrina Alli succinctly describes as "the fabricated crimes of belonging to a surplus population."[12]

Part of what constitutes these "crimes" is the transgression of an explicitly gendered and racialized construction of reproductive respectability. Writing in the context of the incorporation of Black communities into the biopolitics of the US, Grace Kyungwon Hong argues that neoliberalism is accompanied by a redistribution of reproductive respectability and "protectable life." Where racialized subjects were previously exempt from normative modes of reproduction, partial entry into the sphere of reproductive respectability allowed value and vulnerability to be redistributed within and across racialized populations. Access to reproductive respectability, in accordance with the gendered, sexual, and temporal values of cis-heteropatriarchal, familial domesticity, becomes "the dividing line between those who are rendered deviant, immoral, and thus precarious and those whose value to capital has been secured through a variety of norms."[13] This dividing line is the thread that runs through this chapter, distinguishing hardworking families from broken families, *homem de bem* from *favelado*, protected child and from delinquent youth, private resident from homeless itinerant.

Taking into consideration complaints that the application of a revanchist framework "has been, at times, nothing short of Pavlovian,"[14] I draw on a combination of ethnographic accounts, media reports, and political and colonial history to make clear that, rather than referring to an unchanging model of punitive urban policy spreading from "West to rest" or indeed from US to world, auditory revanchism manifests in ways that are geographically distinct. In so doing, I address the continuities and divergences between various neoliberal urban spaces, ranging from deindustrialized regions in North East England to the megacities of Brazil.

I first examine the revanchist logics at play in the use of piped classical music as a deterrent, situating the emergence of this practice in the UK in relation to New Labour's "tough on crime" agenda and drive for a postindustrial "urban renaissance." Second, I address the sonic dimensions of the exportation and adaptation of revanchist logics to São Paulo through an analysis of the deployment and televisual remediation of so-called sound bombs against *pancadões* in the city's poor Black districts. Finally, I draw upon theorizations of "postrevanchism" to examine the introduction of the Northern Territories Anti-Social Behaviour (NTASB) app in Darwin, Australia, which, in enabling users to report those creating noise to authorities, cultivates surveillant listening.

Attending to the conjuncture of revanchist spatial logics and reproductive sound technologies requires an expansion of the latter beyond consumer devices explicitly connected to feminized care and private, domestic settings. Indeed, the sonic surrogates discussed here may initially appear to have little in common with such devices, not least because some do not reproduce sound as sound. Nonetheless, there are coherences: the revanchist reproductive sound technologies described here are also "functional" constellations of mediation, aurality, and affect, bound up with social norms, aspirations, and anxieties about reproduction and contextualized by intersecting social and economic crises centered around housing, homelessness, urban decline, and the breakdown of social and communal relations. Specifically, they speak to the treatment of criminalization and incarceration as one solution to the crisis of reproduction stemming from the rescindment of the waged work and the social wage, and the correlate expansion of surplus populations.[15] They are important insofar as they trouble the "virtuous" character that may be habitually attributed to reproduction and care. As the revanchist criminalization of homelessness exemplifies, carceral institutions place limits upon both normative and alternative methods of reproducing oneself and others. Indeed, carceral and reproductive institutions have become increasingly entangled: Ruth Wilson Gilmore and Craig Gilmore articulate how, in light of the "organized abandonment" accompanying structural adjustments to the welfare state within and beyond the US, social welfare agencies relating to healthcare, social housing, and education have absorbed some of the surveillance and punishment agendas of prisons and policing, while police institutions increasingly participate in the provision and coordination of social services.[16] This chapter thus foregrounds the proximity of care, reproduction, and criminalization as it plays out sonically.

Of particular significance to the sonic surrogates discussed here is the carceral construct of antisocial behavior as a mediator of reproductive respectability. While there are geopolitical variations as to how antisociality is understood and perceived, it is often presented as a confrontation between "proper" and "problem" families. These problem families, which fail to nurture their members adequately, are frequently held to be the responsibility of women.[17] Thus, where the other reproductive sound technologies discussed in this book construe the family household as the primary facilitator of reproduction, revanchist sound technologies aim to step in where the family compact purportedly fails to take care of—and control—its own. As a result, reproductive sound technologies are shown to promote *and police* reproductive respectability.

## "ZERO TOLERANCE WITH A BRITISH FACE": CLASSICAL MUSIC AS DETERRENT

At the beginning of 2023, the BBC reported on a branch of McDonald's in Wrexham, Wales, which had begun broadcasting classical music in response to "bad behaviour." According to the article, twenty to thirty young people had been causing "upset" at the restaurant, leading to the police issuing dispersal orders. Reported incidents included an allegation of assault, children being chased, a fire extinguisher being let off, and coins being thrown at staff. A branch representative states that they implemented the music, alongside other measures, to support the police in tackling such incidents.[18]

The report is one of hundreds of such articles about the use of recorded classical music to regulate the use of shared public and private spaces in UK towns and cities. Indeed, the sheer number of headlines, news reports, and opinion pieces suggests that the significance of this practice partly stems from its ability to generate media attention. After all, the combination of fast food and classical music is rather striking in its incongruity. Yet concealed by this is a notable parallel: classical music—like fast food—is cheap and easy to access. While often associated with high production costs as a form of live performance, classical music has been incorporated into the recording industry from its earliest years.[19] In the twenty-first century, classical music appears as one type of music among many available for low-cost consumption. This apparent cheapness, particularly by comparison to the installation and monitoring of surveillance technologies such as CCTV or the cost of paying for security staff, is one of the reasons that, over the past forty years, broadcasting classical music over public address systems has come to be heralded as a cost-effective solution to the "problem" of antisocial behavior. A BBC report from 2005 titled "Music to Deter Yobs By" makes explicit its cheap expediency: "For the cost of some speakers, and the necessary licence to play piped music, problem areas can apparently be painlessly resolved."[20] Broadcasting classical music is credited with the capacity to dispel, deter, and discourage the presence of "loiterers," an evocative category that refers to many of the "enemies" of urban revanchism, including teenagers, the homeless, drug users, sex workers, and those experiencing mental illness. As a result, and despite the little evidence to support its use as such, piped classical music has been promoted as a crime prevention strategy helping to protect shared social spaces from unwanted and dysfunctional elements.

The use of piped classical music as a deterrent is thought to have begun in 1985, when a branch manager of a 7-Eleven convenience store in the Canadian province of British Colombia began broadcasting "classical" and "easy listening" music into the store's parking lot to prevent teenagers from congregating there.[21] Since then, classical music has been used in North America as a deterrent on public transport systems, in library foyers, and outside restaurants and shops. It has been deployed similarly in the UK: the practice was first reported in 1997, when the Tyne and Wear Metro in the North East of England began broadcasting the English composer Frederick Delius's incidental music from the poetic drama *Hassan* (1923) at some of their stations to target what they referred to as "low level anti-social behaviour," including smoking and swearing. Commenting on its trial at Shiremoor Station in the North Tyneside area, Mike Palmer, the general director of Nexus (brand name of the Tyne and Wear Passenger Transport Executive) stated that the aim was "to provide a background of music that people who we are aiming at don't actually like and so they move away,"[22] describing it as a "win-win" situation: (alleged) troublemakers were driven out, while legitimate passengers found the music helped pass time while waiting for their next train.[23] Following this, Transport for London began broadcasting operatic and instrumental music at forty London Underground stations in 2003. An article on a trial at Elm Park station by *The Independent* describes it as "a place where there was such a gang problem that train drivers were afraid to stop," with the introduction of piped music credited for reintroducing order to the station, resulting in a 33 percent decrease in robberies, a 25 percent decrease in assaults on staff, and a 37 percent decrease in vandalism.[24] The Underground's forty-hour playlist (curated by subcontractor I Like Music) consists of melodic works from the eighteenth and nineteenth century, including works by Handel, Beethoven, Schubert, Vivaldi, Tchaikovsky, Prokofiev, and Liszt. Classical music has since been broadcast in shops, bus stations, restaurants, and bandstands to deter "gangs," "yobs," and "troublemakers." The public body English Heritage (now Historic England) refers to classical music in a 2013 crime prevention guide for heritage assets, suggesting, "where there is an anti-social gathering on a regular basis, consider playing classical music, which may have a calming effect" as such measures can reduce "some of the triggers for illegal and antisocial behaviours" while "contributing towards a more welcoming environment for legitimate [*sic*] users."[25]

As a musical category, "classical music" is both commonly understood and contested, and the resulting definitional ambiguity is discernible from its use as a deterrent. Many media reports cite well-known composers from the Western art music canon: a 2023 press release by the rail operating company Northern, for example, announces that classical music "inspired" by Handel, Mozart, Beethoven, and Tchaikovsky will be piped into 23 stations across their network in a bid to reduce antisocial behavior.[26] Figures associated with Baroque, Classical, and early Romantic music are frequently cited, while references to late Romantic composers and music from the twentieth century are comparatively rare—Nexus's use of Delius being a notable exception. Operatic works and vocal performance also feature: Pavarotti, for example, is mentioned in relation to the Transport for London program. Yet there is little rationale offered as to why particular recordings and compositions are selected and broadcast, with very little comment offered on specific musical qualities. What appears to matter is the affective and social connotations of "classical music."

The initial uptake of piped classical music on the Tyne and Wear Metro occurred around the time that the revanchist rhetoric of "zero tolerance" was being adopted in UK politics by New Labour. In 1995, the future–New Labour Home Secretary Jack Straw gave a speech shortly after visiting New York decrying the "aggressive begging of winos and addicts" and "squeegee merchants who wait at large road junctions to force on reticent motorists their windscreen cleaning service." Graffiti was described as "uncontrolled in its violent image, and correctly gives the impression of a lack of law and order on the streets."[27] During the 1997 election campaign, Straw slightly softened his message of intolerance: speaking in Hartlepool (also in North East England), he claimed that "tackling youth crime" was integral to creating "a more decent society," referring to this approach as "zero tolerance with a British face."[28] New Labour's 1997 manifesto reiterated the party's lack of tolerance for "quality of life" offenses, stating, "We will tackle the unacceptable level of anti-social behaviour and crime on our streets. Our 'zero tolerance' approach will ensure that petty criminality among young offenders is seriously addressed."[29]

New Labour's zero tolerance rhetoric did not lead to the wholesale adoption of New York–style zero tolerance policing in the UK, nor the wholesale adoption of urban revanchism as manifest in US cities. This is partly due to significant economic, ideological, and geographic differences distinguishing Giuliani's and New Labour's political projects. Of note here is the latter's apparent emphasis on social justice, inclusion, and rights, and its substantive *increase* in public spending, albeit mediated by principles of competitive meritocracy (indeed, the New Labour project complicates any notion that the neoliberal welfare state has been steadily shrinking over the course of the late twentieth and twenty-first centuries).[30] Nonetheless, New Labour's deployment of the "zero tolerance" refrain is symptomatic of the internal and external political shifts the Labour party underwent throughout the late eighties and nineties, many of which resonate with the revanchist logics identified by Smith. This included a "tough" approach to crime and order; a preoccupation with the morals and behavior of socially and economically marginalized communities; authoritarian targeting of migrants, refugees, and asylum seekers; and the reorganization of the welfare state such that "benefits" were seen as "a hand-up not a hand-out," conditional on behavior and the ability to become "economically active."[31]

Antisocial behavior became a central construct of the New Labour project: a vaguely defined yet affectively powerful figuration of social and moral decline associated with violent youths; noisy neighbors; drunken yobs; persistent beggars; dysfunctional families; unfit mothers; wayward children; and ultimately, a culture of disrespect, irresponsibility, nuisance, and disorder. As Stuart Hall et al. argue, antisocial behavior was the object of successive postwar moral panics in the UK, and these moral panics engineered support for "increasingly coercive measures on the part of the state" disproportionately impacting poor, working-class, and racialized communities.[32] New Labour both built upon and contributed to this social history, expending significant political capital on the "crisis" of antisocial behavior and the production of new control measures to stem it. Justified as a means of empowering communities and serving the interests of "hard-working families" and the "decent law-abiding majority" and inspired by the "broken windows" approach, New Labour argued that failing to clamp down on such behavior would lead to more serious crime, preventing urban renewal.[33]

In the popular imagination and its political framing, antisocial behavior was strongly associated with "unproductive" poor or working-class young people.[34] It was to be tackled through a range of punitive measures, including fines, dispersal orders, curfews and injunctions. Those found guilty of antisocial behavior could also be subject to welfare restrictions, with, for example, the suspension of housing benefits. Likewise, since antisocial behavior was perceived as a product of "problem" families, those of antisocial young people could be issued with parenting orders, requiring them to attend a guidance program. Addressing antisocial behavior and its associated social ills thus necessitated state interventions into family life and childrearing.

Although not a direct outcome of government or policing policy, the adoption and publicization of classical music as a deterrent in the UK coincides with New Labour's battle against antisocial behavior and quality-of-life offenses. Both are founded on an opposition between a "hard-working majority" and deviant minority, while also relying on ambiguous and shifting notions of crime and criminality. Indeed, although the latter precedes the former, revanchism and antisocial behavior are both conditioned by shifting and expanding notions of criminality that are preemptive and affective. Neil Smith argues that one of the characteristics of the "new" revanchism is "a more active viciousness that attempts to criminalize a whole range of 'behaviour,'"[35] while an expanded notion of crime is "the central marker of the revanchist city, . . . the more so as the fears and realities of crime become desynchronized."[36] Criminality is reframed so that "the sign and symptom become the thing; it is identified with certain forms of social presence in the urban landscape."[37] This is evident in the wording of New Labour's 2003 Anti-Social Behaviour Act which defines antisocial behavior as "behaviour by a person which causes, or is likely to cause harassment, alarm or distress to persons not of the same household as the person."[38] The vagueness of this definition and its reliance on the ear, eye, and affective register of the beholder lends itself to a blurring of preempted and enacted criminality.

While it is often described as a crime deterrent, the deployment of classical music as a tool of dispersal inherits the ambiguous notions of criminality underpinning urban revanchism and antisocial behavior. Little differentiation is made between violent or harmful actions and "loitering" as it pertains to the use of shared or public space in ways that do not conform to commercial interests. The loiterer remains still in spaces where continual movement is encouraged: shops, stations, fast-food outlets, and car parks. They not only fail to partake in economic exchange, they also threaten it: they are held to generate a frightening and menacing atmosphere that repels consumption.[39] To this end, it is revealing that a spokesperson for Nexus claimed in a BBC report that even if those congregating at Tyne and Wear metro stations "didn't have a violent agenda, they looked like they might have."[40] In the same article, a woman commenting on the introduction of "vandal-proof" speakers broadcasting classical music at a bus station in the Yorkshire town of Beverley, suggests that children previously hanging around the bus station after school have been "banished." "The music," she claims "has made a huge difference. . . . I used to feel intimidated by them when I was walking home from work, but now you don't see them standing around anymore."[41] One of the purposes of piped classical music, then, is to produce an affective atmosphere of safety on behalf of those deemed legitimate occupiers of urban space.

As with other forms of programmed music (or Muzak), classical music's use as deterrent relies on the commodification of both music and a listener's affective response to it. There is, however, an apparent tension that accompanies reports of its "audio-affective" functioning, insofar as classical music is understood to both "soothe" and "remove."[42] With regard to the former, classical music has been credited with mitigating deviant behavior through its pacifying influence. The aforementioned English Heritage Crime Prevention guide ascribes it as a "calming" effect, while a 2018 Sky News article on the introduction of classical music to a London branch of McDonald's hails it as creating "an increasingly civil clientele," with a Macdonald's spokesperson crediting the music for resulting in "more acceptable behavior" at the restaurant.[43]

This framing of classical music as encouraging civility and social improvement resonates with long-standing cultural assumptions about its moral value. Many of these are indebted to the historic alignment of classical music and the norms of white middle-class subjectivity, a relationship described by Anna Bull as "contingent but resilient."[44] As she argues, classical music's class formation was initially forged with the industrial bourgeoisie's establishment of music institutions and venues in the 1830s and 1840s, and was further cemented in the 1880s and 1890s through the establishment of key educational institutions and standards. These entrenched sacralized notions of music as art and, by extension, secured classical music's status as a morally virtuous practice. Through the work of social reformers, furthermore, classical music contributed to "rational recreation," aiding in the moral improvement of the working class through directed leisure.[45] To this day, classical music continues to be understood as providing a variety of social benefits due to its moral value and is often presented (and promoted), to use Juliet Hess's phrasing, as "cultural spinach."[46] Classical music, in other words, is "good for you," irrespective of your background, experience, or aesthetic preferences. Such investments have informed various outreach programs intended to help disadvantaged children avoid an "antisocial" life through participation in music.[47]

Although some accounts credit piped classical music with improving listener's behavior in accordance with class-coded notions of respectability and proprietary, it is more commonly described in vengeful terms as an aural irritant working to displace unwanted social groups defined in accordance with age, class, and race. Classical music is considered an effective deterrent insofar as it is assumed that "loiterers" find it intolerable and, as a result, are driven away. This irritating capacity is attributed to both the musical materials and its cultural significations as "elite," "old-fashioned," and "uncool." Nexus's Mike Palmer's description of the Tyne and Wear metro's use of piped classical music highlights this assumed disjuncture, claiming that Delius's music was "the one that would really put the youths off," whereas if the stations had "put on Oasis perhaps we'd gather more youths."[48] Similar social and musical distinctions are drawn in the report in the *Independent* on the London Underground's Elm Park trial: piped classical music is credited with "make[ing] trouble-makers scarper. . . . As for the rest of us, when TfL did a survey of 700 commuters, they overwhelmingly agreed that hearing classical music made them feel happy, less stressed and relaxed." Speculating on the relationship between music's formal properties, mood, and a listener's character, the report concludes that "if calm and stability are alien to you, you probably won't be able to stand Mozart; and your average commuter will probably run a mile if assailed by gangsta rap or Stockhausen en route to the office. Not that hoodies are particularly likely to take to Stockhausen either."[49] The references to "gangsta rap," "gangs," and "hoodies" are telling of how, under New Labour, antisociality was classed as well as racialized: classical music, a white bourgeois formation that affectively benefits "the rest of us," is inevitably repellent to young, "urban" (i.e., Black), and "antisocial" (i.e., unproductive poor) (non)listeners.

While the use of classical music as a deterrent selectively reasserts classical music's "civilizing" potential, it also reflects a decline in its cultural authority.[50] Criticisms of the practice are often concerned with the degradation of a valuable artform: Theodore Gioia, for example, suggests that classical music's use as a deterrent is part of a broader cultural devaluation whereby its secondary associations—generated by advertisements, films, and popular culture—have overtaken its primary meaning. When used as a deterrent, Gioia argues, classical music "devolves from a 'universal language of mankind' reminding all people of their common humanity to a sonic border fence protecting privileged areas from common crowds, telling the plebes in auditory code that "you're not welcome here."[51] While Gioia and other defenders of classical music consider these practices a perversion of its definitive character, the establishment and the policing of boundaries has often been at the heart of classical music: as Anna Bull makes explicit, classical music is one of many projects of symbolic, economic, and geographic boundary drawing that typify white middle-class subjectivity. Like suburbanization, private education, and gated communities, classical music has been and continues to be a site of class-based, gendered, and racial boundary-making through which the middle classes reproduce themselves. These boundaries are manifest across its institutions, educational curricula, and modes of social organization: from the sealed space of the concert hall aurally cut off from its urban surroundings to ideas of musical propriety, seriousness, and hierarchies of ability.[52] Recognizing classical music's practices of boundary work thus offers an alternative means of accounting for its use as a revanchist sound technology and its seemingly contradictory abilities to soothe and remove: rather than representing a substantive break with its prior valuation of art, its adoption as an audio-affective deterrent marks a continuation of classical music's dual status as a civilizing technology of self-improvement and a boundary-making technology of exclusion.

Classical music's boundary work is not only symbolic. Marianna Ritchey has powerfully articulated how classical music's institutions materially participate in urban gentrification and the accompanying intensification of class-based and racial stratifications. Classical music's buildings have been used to attract private investment to currently "nonlucrative" poor and racialized neighborhoods, a dynamic that ultimately contributes to processes of urban displacement. Meanwhile, she notes, representatives from finance, banking, local government, and property development frequently feature as donors, partners, and board members of orchestras and concert halls, leading to a situation in which some of those funding and directing music institutions, including their inclusion and outreach programs, are themselves beneficiaries of punitive, exclusionary, and racist forms of urban regeneration.[53] Although Ritchey is writing in a US context, some comparable dynamics are at play in the UK, where music institutions have also acted as drivers of urban "regeneration," directing capital (via property developers, landlords, and business) to revive "declining" neighborhoods. To this end, it is striking that the use of classical music as a revanchist sound technology in the UK seemingly originates in the North East of England, and at the same time that the region was undergoing another form of musicalized renewal.

Over the course of the twentieth century, the North East was significantly impacted by deindustrialization and its social fallout. New Labour's election victory in 1997 was underpinned by a promise to restore prosperity to such areas, which were the party's historic heartlands.[54] One method through which New Labour sought to achieve this was "creative regeneration." At the same time Delius's music from *Hassan* could be heard at Shiremoor station, plans were underway to transform a brownfield site a few stops away in Gateshead into a world-class music venue that would provide a home for the chamber orchestra Northern Sinfonia. Occupying an area of quayside depopulated by the 1930s and 1960s slum clearances, The Sage Gateshead (renamed in recent years The Glasshouse Centre for Music) opened its doors in 2004, two years after its near neighbor, Baltic Centre for Contemporary Art. Surrounded by newly built residential buildings and hotels, these "flagship" cultural developments, and the development of the quayside more broadly, were the result of substantial public funding. However, one of their purposes was to attract private investment by reconfiguring the area's identity to one of tourism-amenable cultural vibrancy. According to government figures, the £250 million of investment by Gateshead Council into its arts, leisure, and residential redevelopment generated over £1 billion in private investment.[55] The Sage itself was named after one of its donors, the multinational software company Sage Group; while one of its performance spaces is named after Northern Rock Foundation, a charity attached to the British bank Northern Rock, which collapsed during the 2007 subprime mortgage crisis.

The development of the quayside coincided with another regeneration scheme involving public funding, "Bridging NewcastleGateshead," which was part of New Labour's Housing Market Renewal (HMR) Pathfinder program launched in 2002. HMR schemes directed public and private investment into deprived urban neighborhoods that were understood to be at risk of "market failure" due to low housing demand. Through a mixture of demolition, construction, and refurbishment, HMR schemes sought to "modernize" the housing stock, enabling the participating neighborhoods to attract higher-income residents and to reintegrate them into the regional property market.

Bridging NewcastleGateshead (BNG) covered innercity neighborhoods in East and West Newcastle, as well as those close to the river in Gateshead. While the connections between BNG and the cultural regeneration of the Quayside should not be overstated, there were intersections between these two urban development projects. In its accompanying literature, BNG refer to the quayside and its cultural institutions as signifiers of wealth and success, with which the target neighborhoods need to be aligned: its reported aim was to "extend the revisioning of the waterfront—the Sage, Baltic and Quayside—eastwards to Walker Riverside."[56] Even the name "Walker Riverside" is an instructive "rebranding" of the deprived Walker area to the east of Newcastle city center, coinciding with the quayside regeneration projects.[57] That BNG envisages itself as expanding the success of the latter is perhaps unsurprising given the involvement of Gateshead Council and the regional development organization One North East in both projects. Furthermore, indicators of gentrification—the decline of social housing and the expansion of the private rental market and owner occupancy, rising rates of speculative property investment and rapid increases in rents and property prices—were more pronounced in Gateshead's regenerated neighborhoods close to the Tyne, implying that the cultural revitalization of the Quayside was contributing to the area's social and economic reconfiguration and, by extension, dynamics of urban displacement.[58]

The piped music at Shiremoor station and the development of the Sage/Glasshouse resonate with two distinct but connected aspects of New Labour's neoliberal urbanism: a punitive, moralistic social neoconservatism preoccupied with antisocial behavior and the cultural and economic threat of the failed family form; and postindustrial renewal involving substantive public (and private) investment in housing and culture. What connects these is the threat and the "promise" of displacement. Between Shiremoor and Gateshead, classical music appears as a tool of revanchism capable of driving out the antisocial elements of urban environments and as a tool of gentrification, contributing to the reconfiguration of the cityscape to attract investment, business, and the professional middle classes. This second claim is undoubtedly more controversial and requires caveats. The Sage/Glasshouse places significant emphasis on its outreach, learning, and participation programs, and the pricing out of the urban poor and working class is likely antithetical to the interests of many who work at the institution. It would also be a mistake to assume that the drive for gentrification by some stakeholders has been fully realized: the region's development has remained uneven. Nevertheless, in contributing to efforts to attract private investment and "uplift" housing markets, the Sage/Glasshouse is proximate to the drive to bring capital back to the city.

## CLEANSING THE SOUNDSCAPE: HYGIENIZATION, *PANCADÕES*, AND SOUND BOMBS

Although many analyses of urban revanchism have been empirically grounded in North America and Western Europe, some scholars have presented processes of urban regeneration and displacement as global or "generalized" phenomena. This has not simply involved the transmission of urban policy from North to South or along a core to periphery trajectory. Rather, as Loretta Lees, Hyun Bang Shin, and Ernesto López-Morale suggest, policies and practices of urban renewal and displacement are multicentered, and have long been manifest outside Western Europe and North America.[59]

In Brazil, the zero tolerance approach of Giuliani (who has been hired as a security consultant by a number of Brazil's federative states since at least 2009) is combined with localized racial-spatial formations to produce a distinctive form of revanchism. As Jeff Garmany and Matthew Richmond make clear, *higienização* (hygienization), pertains to a racial and colonial history that conflates modernity, cleanliness, and whiteness, and which informs dynamics of urban displacement—and importantly, *containment*—in Brazil. Following the abolition of slavery in 1888, Brazil's newly formed republican government sought to resist the perception of the nation as a backwards or underdeveloped country through the pursuit of a modernizing agenda. Inspired by eugenicist theories linking modernization with genetics, they (and other proponents of modernization) sought to "whiten" the population by incentivizing immigration from Europe and prioritizing white immigrant access to land and labor markets.[60] These principles were also applied to the management of cities, where architects and planners emphasized the importance of moral and social hygiene in their work. This involved clearing areas with high population densities and poor public sanitation, in effect diminishing the means of poor and Black urban survival. Not only were these areas deemed unhygienic and hazardous to the health of the population, they were also signifiers of an underdeveloped and unenlightened past. Consequently, the adoption of hygienization in postcolonial Brazil concerns the defense of progressive (white, European) civilization from the unproductive, and racialized masses.[61]

A key social legacy of hygienization is the favela, and its treatment as a hazardous site of contagion that requires quarantine and, ultimately, eradication.[62] As Garmany and Richmond note, this differs from European and North American processes of gentrification, in that the primary motivation is not necessarily the maximization of rent or investment opportunity but the imposition or restoration of "hygienic urban landscapes," often accompanied by direct forms of state violence and the pathologization of the low-income residents as delinquent trespassers. At the same time, the desire for "order" and accumulation often converge, particularly as Brazilian urban life is increasingly structured around financialization.

In December 2006, São Paulo's State Secretariat for Public Safety issued a statement celebrating the success of a new policing program. Among the most populous cities of the world, São Paulo is the financial and economic center of Brazil. It is also a highly unequal and geographically segregated city: among other factors, racialized structural unemployment, white supremacist prohibitions on property ownership and urban zoning, internal migration, low and disparate wages, patchy welfare provision, high land values, waves of urban renewal and the return of the middle classes to the city center, all conspire to produce an urban order that combines a spacious, securitized playground for wealthy elites. Meanwhile, the poor and Black (the latter of which provide a cheap labor force frequently exploited by white households in the form of domestic service) live in precarious, overcrowded, and illegal compounds,, which house much of the city's surplus population.[63] Indeed, São Paulo's urban formation is both multiracial and anti-Black. This (oft-denied) anti-Blackness is clearly traceable in the city's history of law enforcement, and in disproportionate rates of incarceration and police killings it has produced.[64]

In 2006, however, the state secretariat was hailing a different, more effective approach to securing the city, drawing a direct comparison between the new measures in São Paulo and Giuliani's interventions in New York:

> São Paulo's effort to reduce homicides is evident when comparing the results of the city with the ones obtained by the known New York "zero tolerance" program. The fall in the percentage of homicides in the third trimester of the last seven years (1999 to 2006) was 64.14 percent, overcoming the percentage achieved by New York during Rudolph W. Giuliani's term.[65]

Since the late 1990s, state forces in São Paulo have sought to reduce the city's high rates of crime while also cleaning up the city's image as a violent metropolis. Attracting urban investment, international businesses, and tourists required São Paulo to loosen its associations with both crime *and* state-sanctioned killing. Hence a new policing program was introduced: whereas the "old" police were associated with human rights abuses and extrajudicial killings, a series of "progressive" security strategies were established targeting the city's low-income, multiracial periphery. These included encouraging local residents' participation in neighborhood security councils, human rights training for police officers, anticrime campaigns in schools, collaborations with NGOs and Christian missions against drugs and domestic violence, and the implementation of "community" policing. These strategies imply a shift from a militarized and repressive mode of policing towards a "softer" model of governance.[66]

As the state secretariat's 2006 public statement intimates, a component of São Paulo's new policing is a zero tolerance approach, with a focus on minor misdemeanors, antisocial behavior, and drug related crimes.[67] In Brazil, then, the establishment of zero tolerance, follows a different pattern to that of revanchist New York, appearing as a market-friendly alternative to the overtly repressive violence of the "old" police. Put differently, zero tolerance in São Paulo does not mark a newly punitive and vengeful approach to the urban poor, who were already treated and policed as enemies of the city. Rather, as Neil Smith, notes, the adoption of the language of zero tolerance in São Paulo "gives a professional and pseudo-scientific alibi to a far more brutal and militarized revanchism than yet occurs in New York."[68] The new policing strategies did not simply replace the old:

rather, as Jaime Amparo Alves emphasizes, they existed alongside and are connected to the persistence of anti-Black state necropractices: the racial-spatial containments, mass incarcerations, family separations, and police killings integral to the reproduction of Brazilian social order and, with them, the biological and economic prohibition of reproducible Black life. Thus, "at the same time government and think-tank discourses sold the city as a security project that required the active participation of its citizens, shoot-to-kill policies and the incarceration of the predominantly black (surplus) population intensified."[69]

The imbrication of anti-Blackness, hygienization, revanchism, and São Paulo's seemingly divergent methods of policing is made apparent by state responses to *baile funk* street parties. 2012 saw the launch of "Operation Pancadão," a collaboration between São Paulo's military police and the *Programa de Silencio Urbano* (the urban silence program, or PSIU), a law enforcement agency that addresses noise complaints, consisting of a set of actions intended to combat noise and prohibit the consumption of alcohol by underage drinkers. Although the PSIU may be called to address noise-related issues coming from a variety of sources, including commercial venues, churches, and construction sites, the naming of Operation Pancadão invokes a specific auditory target.[70]

In 2011, *pancadão* began being used in the Brazilian media to name a type of loud and unregulated street party popular in São Paulo's poor and lower-middle class areas. Roughly translated as "a hard hit," the term originates with the *funk carioca* scene, where it refers to a style of beat. A bass-heavy style of electronic dance music combining elements of Brazilian and North American musical styles, including Miami bass, hip-hop, samba, and capoeira, *funk carioca* stemmed from Rio de Janeiro, where it became popular among predominantly Black Brazilian favela communities in the 1980s and early 1990s. Although it has since gained international acclaim and become an important musical export, *funk carioca* has also been subject to legal contestation domestically. Exemplifying Brazil's hygienist racial-spatial politics, *funk* parties have been framed as a threat to public health and a site of contagion: politicians, the police, and the media have, with popular support, framed "the sound of the favela" as spreading violence, criminality, and transgressive sexuality. Licenced *funk* parties may be permitted and have previously received funding via the secretary of culture, but this requires organizers to work with institutional and state agencies, including the police.[71]

In São Paulo, *pancadãoes* are closely associated with a local variant of funk music: *funk ostentação* (ostentation funk). Emerging in 2008, it departs from *funk carioca*'s tropes of sex, violence, and social justice, and instead places emphasis on conspicuous consumption. The music's lyrics and visual aesthetics often emphasize symbols of economic and social success: money, luxury goods, women, alcohol consumption, and cars. Frequent musical and visual references to the latter are mirrored by the use of cars as a sound system during *pancadãoes*, which often involve multiple customized cars with large built-in speakers simultaneously playing *funk ostentação*. Leonardo Cardoso describes this imbrication of musical content and performance as a mode of "sonic ostentation": "the music that talks about ostentation comes from an ostentatious device (the audio equipment), which is integrated into another ostentatious device (the vehicle)—a triple ostentatious move."[72] However, the car's significance to the *pancadão* isn't only symbolic or acoustic: it also helps ensure the longevity of the party, enabling it to move to a different site if and when the police arrive.

While ostentation has long been a marker of power within Brazilian society, some commentators have associated *funk ostentação* with expanding economic prosperity amongst the urban peripheries, which have enabled greater access to consumption.[73] Others have suggested that *funk ostentação* is symptomatic of a political emphasis on ostentation over welfare. The Rio de Janeiro artist MC Koringa, for example, has emphasized the connection between ostentation in the favela and a lack of meaningful investment in infrastructure and education for the urban poor:

> There is a cable car, but waste collection is deficient. The schools are not always open, so you produce ignorance. Transportation systems are political make-up. From slaveships to the metro today, it is the same thing.... Instead of putting up a cable car on a hill, you need to solve the problems down on the ground. What good is a cable car if you don't have the security to use it? Ostentation [in the favelas] is the mirror of these failures of the government.[74]

However, the exertion of the neoliberal right to consumption by some of São Paulo's poor and working classes has been met with racial and class-based anxieties among the city's wealthier residents. In addition to the *pancadão*, *funk ostentação* has been associated with another youth practice: *rolezinho* (roughly translated as "a little stroll"). During 2013–2014, large groups of young people from poor and working-class neighborhoods began using social media to organize large-scale gatherings at shopping malls, where they would sing, dance, and walk around. The first of São Paulo's *rolezinhos* was held at Shopping Metrô Itaquera to the city's east: six thousand young people showed up after it was advertised on Facebook. Many of them were reported to be *funk ostentação* fans, dressed in designer baseball caps, sunglasses and rings.[75] Unsurprisingly, *rolezinhos* were deemed an antisocial threat and quickly became subject to a security crackdown.

Like *rolezinhos*, *pancadãoes* are involved in a racial-spatial conflict over leisure time. For some, they are an opportunity for young people with limited access to urban leisure space to dance, drink alcohol, flirt, renew social bonds, and challenge the police away from teachers, bosses, and parents.[76] For others, including the police and some attendees of neighborhood security councils, they are sites of noisy disorder threatening reproductive respectability and endangering its conduits: children, families, and workers.[77] To this end, the reception of these parties affirms a social distinction between criminals and *homens de bem* (good men, or men of good will): what Alves describes as a moral category, mainly comprising of white or light-skinned local businessmen, NGO staff, employees of politicians, and churchgoers. While *homens de bem* live in São Paulo's low-income districts, they have access to—and are willing to work with or advocate for—politicians and police. They are the main attendees of community security councils and are threatened by their proximity to the "disorderly" space of the favela, associated with the purportedly uneducated, politically illiterate and antisocial resident of "illegal" settlements, who is often believed to be associated with the underground drug economy. These contrasting figures of the good man and the criminal, the community-oriented citizen and the antisocial deviant, are invoked by the Brazilian police, conservative media, and participants in local security councils.[78]

Implicated in the opposition that Alves identifies is a distinction based on work ethic and participation in formal economies. The *homem de bem* tends to be a worker—specifically, a hard worker. The criminal *favelado*, by comparison, exists apart from the world of formal employment and its temporal rhythms. In Alves's account, this distinction is captured in the comments of an attendee of a community security council in the south of the city:

> He complained that "the guys" [drug dealers] party all night during the weekend, and even during the week they keep loudspeakers on the streets until late at night. He requested that the police close the bars and confiscate the speakers because the bars are places for drug sales. "These are not *pessoas de bem* [good people]" he explained. "A good citizen does not stay out late partying during the week."[79]

Not only does the criminal *favelado* fail to conform to the rhythms of work and rest; their loudspeakers and funk parties are actively disruptive to them. The noise of the *pancadão* inhibits the sleep of the *homens de bem*, continuing through the night both at weekends and during the week. Consequently, police responses to the *pancadão* protect the peace and quiet of "worker" residents so they can go about their jobs the next day.[80]

Informed by the distinction between hardworking *homem de bem* and the criminal *favelado*, the *pancadão* is a site of racial anxieties regarding sexuality and sexual reproduction. A recurring figure in state and community security council responses to *pancadãoes* is the pregnant, or pregnable, young woman. The *pancadão* has been met with an anti-Black (anti)reproductive politics, articulated in fears about the teenage girls in attendance participating in sexual activities and becoming pregnant by "criminals."[81] Given the association of Blackness with both *baile funk* and state-defined notions of the criminal, concerns about *pancadão*-related pregnancies are bound up with wider hygienist anxieties about the racial constitution of the urban population: the figure of the pregnant or pregnable *pancadão* attendee is a threat to white-constituted social order insofar as they contribute to the production of these "criminal" or "unproductive" elements of the population.

The attendance of actually pregnant women at *pancadãoes* has been used as further evidence of the threat they present to hygienist notions of reproductive respectability. A 2015 video, *polícia acaba com pancadão* ("police end a pancadão"), originally aired on the police journalism television show *Operação de Risco* (*Risky Operation*), documents a raid on a party in an undisclosed neighborhood. The program's host describes the military police as imposing an "orderly shock" to a prohibited party, attended by "mostly criminals."[82] As Pedro Vieira de Oliveira notes, despite the use of dramatic music and language in the video, the operation carries little risk for the police. Instead, "the video showcases a disproportionate use of military technology, as an excuse to, in the words of the officers '[protect] the neighbors from noise, so they can rest and work tomorrow.'"[83] Emulating the audiovisual aesthetics of first-person shooter games, the police are shown clearing a party in a bar. Among those exiting the building is a visibly pregnant woman, wearing a tight crop top and low-rise jeans. Her appearance in the video is accompanied in the soundtrack by a dramatic percussive "sting," while an officer accusatorily asks her, "What the hell are you doing here? You should be at home!"[84] The focus on the pregnant attendee in this sensationalized remediation serves to affirm the "amorality" of the *pancadão*, as a site in which reproductive respectability is breached. From the perspective of security councils, the police, and sympathetic viewers of *Operação de Risco,* the repression of the *pancadão* is thus heard as both enabling and inhibiting reproduction: it preserves the daily reproduction of workers and the reproduction of the social order while preventing teenage pregnancies, and with them, the reproduction of an unproductive Blackness.

Although considered an act, site, and source of criminality, Leonardo Cardoso suggests that *pancadãoes* are somewhat resistant to policing due to their number (in 2012, police reported 300 parties occurring a night) and the mobility afforded by their car-mounted speakers. Despite many nearby residents calling the police to complain about the loudness of the events, Cardoso suggests that "both police and partygoers seem aware that it is almost impossible to repress these parties with physical force."[85] Nonetheless, as the *Operação de Risco* video attests, *pancadãoes* have been subject to violent repression by the police. This has included the use of sonic violence through the deployment of military technologies such as explosive stun grenades or "sound bombs." Indeed, these have a starring role in the video: the police are shown firing a grenade launcher into a crowd, generating a loud explosion that is quickly followed by the sound of car alarms and screaming. As Vieira de Oliveira observes, the sound bombs used by the Brazilian Military Police were developed by the US Army as "clear-a-space" devices in the 1990s: they create a loud bang of up to 180 decibels at fewer than two meters so as to disorient or debilitate a target. Although deemed nonlethal, they can cause significant harm, including hearing damage and shrapnel-related injuries from the grenade's metal canister.[86] Indeed, the relative "safety" of sonic weapons typically concerns their users rather than their targets, as alluded to during the *Operação de Risco* video: the sound bomb is described as guaranteeing a "safe course" in the operation, insofar as it mitigates physical confrontation between civilians and the police.[87] To this end, sound serves as a surrogate for physical force, minimizing the need for riskier and more labor-intensive police tactics.

Deployed in the context of Operation Pancadão and *Programa de Silencio Urbano*, the military police's sound bombs constitute a cleansing of the soundscape and a silencing of a musical life deemed antithetical to the modernizing drive for investment and accumulation in São Paulo. The use of these devices and their spectacular remediation is enmeshed with two notable contradictions. On the sonic level, the orderly peace and quiet that enables "rest and work" is generated through extremely loud and disruptive sound of both the sound bomb and its secondary auditory events (i.e., screaming and car alarms). For the video's intended audience, sound bombs are presented as an expression of police power and efficiency, yet the use of militarized technologies in a context in which the only depicted evidence of criminality is a private party in a bar seems strikingly disproportionate. The use of sound bombs is justified by appealing to the vulnerability of the police (and by extension the "good" community): sound bombs help to ensure police and community safety, by harnessing noise to reassert state control in response to the auditory, spatial, and social transgressions of the *pancadão*. In this regard, the sound bomb acts as a revanchist reproductive sound technology—a means of protecting the designated temporalities and spatialities of work, leisure, and rest from the "insurgent" sounds of racialized surplus life.

## NOISE, POSTREVANCHISM, AND INTRA-INDIGENOUS CARE

In 2018, the Australian Northern Territory Government announced a new suite of initiatives to tackle an apparent protracted crisis of antisocial behavior, of which, in the words of the Chief Minister of the Northern Territory Michael Gunner, "Territorians have had a gutful."[88] Accordingly, a trial of the "NT Anti-Social Behaviour" (henceforth NTASB) mobile app was launched in 2019, enabling residents to effectively report disruptive, concerning, or damaging activities in public spaces in Darwin and Palmerston. Marking an incident by dropping a pin on an app, users could select one of four options to characterize the antisocial behavior in question: public drunkenness/intoxication, sleeping, camping and homelessness, welfare check, and noise or nuisance. Accompanied by an icon of a human figure with sound waves emitting from their head and surrounded by musical notes, examples of behavior to be reported under the latter included playing of loud music in a public space, using offensive language, begging, and "humbugging."[89]

To present the NTASB app as a reproductive *sound* technology may appear a stretch: unlike the majority of devices and services that have been discussed under this rubric thus far, it does not produce or reproduce sound aurally but rather logs complaints about unwelcome or distressing behavior. However, the presentation of noise as a mode of antisociality reveals an aspiration to intervene in auditory relations. Like the feminized smart speaker examined in chapter 2, NTASB app is predicated upon surveillant listening. Where the smart speaker's surveillant listening appears as a service that benefits the user, here surveillant listening is collectivized and crowd-sourced, ostensibly to the benefit of resident citizens. To this end, the NTASB app is a type of recording device. What is recorded, however, is not audio but aural experiences deemed by a "possessive" listener as unwanted and disorderly.

The trial of the NTASB app was met with criticism and concern. Articles and social media comments suggested the app and the government were criminalizing homelessness, and in so doing, targeting Aboriginal communities in ways resonant with (enduring) colonial and postcolonial histories of displacement, dispossession, surveillance, and subjugation that serve to, in the words of Aileen Moreton-Robinson, reproduce the Australian nation-state as a "white possession."[90] As Moreton-Robinson argues, Australian neoliberal discourses recognizing formal equality through citizenship and stipulating that the market determine the life outcomes of human beings are integral to both the prohibition of Indigenous self-determination and the maintenance of Indigenous "pathology," "enabling the impoverished conditions under which Indigenous people live to be rationalized as a product of dysfunctional cultural traditions and individual bad behavior."[91] Relatedly, Indigenous communities in and beyond the Northern Territory have long been subject to overtly revanchist attitudes. In 1996, the Lord Mayor of Darwin described the city's approach to Aboriginal "deviants" as follows: "Harass, harass, harass. . . . if you keep shifting them around, constantly harass them so they can't settle, they will get sick and tired of it and maybe some of them will go back to their own communities."[92] As Chris Howse notes, such statements "shine a quite pitiless light on a dynamic of long standing in the Northern Territory: that people in government believe . . . they must humiliate Aboriginal people to control them."[93] Consequently, Aboriginal communities have been disproportionately targeted by "move-on powers" enabling police to instruct individuals or groups to leave a public space in cases where they are (believed to be) about to commit an offense, are creating an obstruction, or are creating "anxiety" for others around them.[94] More generally, Indigenous people are overrepresented within the Australian criminal justice system, and are more likely to die in police custody.[95]

One effect of these rates of incarceration is the further disruption and fragmentation of Aboriginal kinship relations by the Australian settler state. Since colonization, government policies, laws, and legislations have routinely targeted Aboriginal families: the forced removal of Aboriginal children from their families and communities in accordance with "assimilation" and "protection" policies from the beginning of the twentieth-century is emblematic of state hostility and suspicion toward Aboriginal social life.[96] While such policies were superseded in the 1960s, the notion that Aboriginal families are dysfunctional, neglectful, and harmful and thus require intervention has endured. The government's response to a 2007 Northern Territory Board report into the protection of Aboriginal children from sexual abuse, for example, was to implement a range of punitive policies including banning alcohol, gambling, and pornography from Indigenous communities in the Northern Territory and placing restrictions on how welfare payments could be spent. Given this charged history, it is unsurprising that the NTASB app appeared as another instance of the settler state seeking to survey, police, and intervene in Indigenous life.

The app, and criticisms of it, also reflect issues around housing and homelessness in Darwin and the Northern Territory. A small and isolated coastal city (as of 2024, the estimated resident population of Greater Darwin was 152,489[97]), Darwin commands high property and rental prices. These are accompanied by high levels of homelessness: the 2018 report of the government-funded Australian Homelessness Monitor notes that between 2011–2016, Darwin had the highest rate of homelessness of any Australian capital city, while the Northern Territory has the highest rates of homelessness in Australia.[98] Writing in 2013, Parsell and Phillips suggest that Indigenous people make up 95 to 99 percent of Darwin's large and highly visible rough sleeping population, who "live in the long grass" on a temporary or longer-term basis.[99]

The city's "long grass campers" or "long grassers" are a community, or set of communities, residing in nearby bushland, as well as parks and streets, some of whom construct shelters with distinct areas for sleeping, cooking, and washing. Various individual, cultural, and structural factors that draw people to long grass camping. Some arrive in response to experiencing overcrowded and inadequate housing in the Northern Territory's remote communities, after fleeing violent and abusive domestic situations, or in order to evade the oversight of the settler state (including forms of "care" experienced as dispossession or incarceration); others choose to stay in the long grass communities while visiting friends and family, accessing the city's health and education services, attending court dates, or attending cultural and sporting events. Some occasionally sleep in the long grass but mostly reside elsewhere in Darwin, while others stay for longer periods.[100] To this end, the long grass does not simply provide a space to sleep and socialize but is a site of broader domestic and reproductive activity. The Wiradjuri archaeologist Kellie Pollard notes that many of those living in the long grass participate in self-care practices directed at maintaining their health, hygiene and personal appearance, and prepare both shop bought and hunted bushfoods.[101] In the long grass, then, "homelessness" frames a heterogeneous collection of living and sleeping practices.

Long grass camping has long been stigmatized by settler-Australians and its media culture as a mode of antisocial behavior, performed by an abject and criminal underclass. In the context of Darwin's expanding tourism economy and property market, furthermore, campers are held to be bad for business and asset values in the region. As Daniel Fisher argues, the production of long grass camping as a media spectacle in the early aughts was "increasingly at odds with Darwin's new self-image as a gentrifying, property conscious, northern destination: welcoming to tourists, safely multicultural, on the doorstep of Aboriginal Australia, and a gateway to Australia's natural wonders at Kakadu national park."[102] As a result, substantive resources have been directed toward the removal of campers from public spaces, with critics of these policies responding by figuring campers as displaced and homeless people or victims of trauma and violence who, in their attempt to evade misguided state interventions into their personal and family lives, find themselves exposed to further harm.[103] As will become clear, such articulations of long grass camping are themselves revealing of a range of institutional and political motivations.

The NTASB app affirms the framing of long grass camping as a mode of antisocial behavior producing those living and sleeping in public spaces as a social and economic problem requiring reporting and intervention. Although an update rolled out in 2021 replaced the option to report "sleeping, camping and homelessness" with "obstructing public access," it still included the example of "camping/living in unauthorized areas." As a result, the app reimposes a private/public distinction that is central to settler order but which exists in tension with Indigenous modes of life. As Pollard et al. argue, Aboriginal people have camped in the long grass since the beginnings of European colonization and so the use of "public" space in this way is a continuation of cultural practice: "the response of authorities to Aboriginal homelessness in the long grass is a denial of Indigenous agency, culture, and rights to country."[104] Furthermore, the app associates those living and sleeping outdoors with other types of nuisance and criminal behavior, including the making noise, which the app describes as impinging on the right of "residents" to live free from public disturbance.

The notion of noise has long functioned as a tool of marginalization, operating as a means of producing certain groups as disruptive, disorderly, excessive, out of place, or threatening to social norms within urban environments. The pejorative association of noise with minoritized populations traverses its various legislative, aural, and informational definitions. In North America, Western Europe, and Australia, complaints about urban noise and attempts to mitigate it have often attributed responsibility to poor and working-class communities and to racialized, colonized, and immigrant populations. In her examination of New York City's antinoise legislation in the early twentieth century, Emily Thompson makes clear that those targeted as noisemakers were relatively powerless populations "who impeded, in ways not just acoustical, the middle-class vision of a well-ordered city."[105] The employment of noise ordnances to clamp down on the activities of street vendors, all the while leaving the noise of traffic unregulated, demonstrates how noise bans ultimately functioned as a means of clearing the streets to make way for a productivist—and even louder—modernity.[106] Similarly, in his reading of the English polymath Charles Babbage's 1864 pamphlet on London's street musicians, Seb Franklin highlights the conjuncture between informational understandings of noise and the attribution of discordant and excessive sound to the surplus populations. For Franklin, those who are deemed corrosively "formless" due to their deviance from value-mediated social relations mirror information theory's notion of noise: just as the surplus population is simultaneously produced by, necessary to, and externalized by capital, so too is noise produced by, necessary to, and externalized by communication channels. In making this connection, Franklin highlights Babbage's exasperated account of the (supposedly) detrimental social and economic effects of street musicians, whom he saw as promoting pathological sexuality, outlawed forms of reproduction, and idle frivolity and as threatening property values and the productive activities of the intellectual and industrious worker.

In more recent noise abatement campaigns, noise has continued to constitute unproductive social deviants. In their study of the online antinoise community group Quiet Australia, established in 2003, Buck Clifford Rosenberg observes that members' posts generate a distinction between quiet, good, self-regulating citizens and bad, noisy, uncivilized, and uncaring others who require state intervention and regulation. Where some posters are relatively measured in their suggestions for dealing with neighbor noise issues, others resort to hyperbole and dehumanization to articulate the threat of noise faced by "decent" residents. One poster, for example, complained of noisy neighbors who insisted on "talking outside on mobile phones," "playing loud music," and having "shouting conversations," and who have "barking dogs" and "noisy children." In response, another poster characterized this as "a manifestation of the disintegration of Western society where selfishness rules supreme": "why should decent people keep on suffering because of the garbage breeding in our society[?]"[107]

In encouraging users to participate in racially marked practices of aural surveillance by reporting antisocial noise created by those that socialize, sleep, and live in (settler-defined) public spaces and in presenting this noise as an impingement on the rights of "residents" that requires intervention, the NTASB app echoes the auditory revanchism of Guiliani's New York. Indeed, the app represents the prevalence of what Chris Herring calls "complaint-oriented policing" involved in "quality of life" enforcement, where (propertied) citizen's complaints drive interactions aimed at "*neutralizing poverty* through incapacitation and invisibilization."[108] Again, however, there are important distinctions between New York's punitive, reactionary, and vengeful policing of sounds and activities and the methods of intervention offered by the NTASB app. Where the former is firmly grounded in a criminal justice framework, for which "disorderly" behavior is a mode of criminality; the latter is grounded in a more complex and sometimes contradictory politics of social control that both incorporates *and* works against carceral approaches to social "problems" involving Indigenous communities.

Although it has been presented and promoted as part of the Northern Territory Government's five-point plan for tackling antisocial behavior, the NTASB app was developed and runs in partnership with representatives of the Larrakia: an Aboriginal nation, presently consisting of around 2,000 people, who claim traditional ownership of the Darwin region.[109] Larrakia economic, cultural, legal interests are represented by two Aboriginal corporations: the Larrakia Development Corporation and Larrakia Nation Aboriginal Corporation. The former seeks to provide economic opportunity and support by facilitating land development and business, employment, and investment initiatives, with a portion of profits from these commercial activities providing financial assistance for Larrakia families.[110] Founded in 1997 to allow a corporate identity for Larrakia people looking to uphold Native Title claims, the Larrakia Nation Aboriginal Corporation provides cultural and legal representation, as well as a variety of community and outreach services. Primarily targeted at "Aboriginal visitors and homeless people" to provide "early intervention to divert people away from dangerous situations, contact with the criminal justice system and link them with support services such as healthcare," these services include a night and day patrol and a health and welfare assistance program for those staying in the long grass.[111] The NTASB app intersects with these services. As a result, it is a (partial) expression of the Larrakia's authority as traditional landowners, and an intra-Indigenous politics of care.

In contrast to settler framings of the long grass campers in terms of criminality, the Larrakia Nation Aboriginal Corporation position these "homeless" populations as in need of culturally appropriate care. To this end, the Corporation frame the NTASB app as enabling culturally appropriate care to be administered. If a report is made through the app, the Larrakia Nation patrol service is notified. According to the app's FAQ page, patrol officers assist by defusing situations in which antisocial behavior is taking place. This can involve relocating a person to a "safe" environment (such as a relative's home, refuge, or hospital); diffusing violent situations; providing advice, information, or referral to other services; and diverting intoxicated people from contact with the criminal justice system.[112] In 2021, Larrakia Nation suggested they were receiving around 300 reports a month through the app and expected demand to increase as the service became increasingly publicized.[113]

Despite being centered around the reporting and mitigation of antisocial behavior, then, the app is nonetheless informed by an Indigenous, noncarceral approach to harm reduction that is grounded in mediation, negotiation, and service provision. To this end, Larrakia Nation seek to work against normative approaches to "quality of life" enforcement by distinguishing antisocial behavior from crime. In a news post on the organization website, Russel Wills, a Northern Territory civil servant, who was assigned to the project and managed the development and the rollout of the app, notes that it is "important that people realise that while antisocial behaviour is unacceptable it is distinct from crime and is treated differently."[114] Yet the app itself does not always make this distinction described by Wills clear: under the section on "excessive noise and nuisance," for example, users are instructed to call for emergency assistance, using 000, if there is “excessive noise [*sic*], fighting, or violence of any kind.”[115] As a result, it is left ambiguous as to whether “excessive noise” constitutes antisocial behavior benefiting from a noncarceral approach or a criminal activity necessitating police intervention.[116]

The NTASB app’s conflicting politics of criminalization and care are arguably symptomatic of wider “postrevanchist” tendencies manifest in some urban environments, where policies concerning homelessness, antisocial behavior, and the use of public space tend to be informed by a more ambivalent, softer, or “compassionate” approach than the “zero tolerance” rhetoric exemplified by Giuliani’s 1990s New York. The emergence of a postrevanchist urbanism has meant that these punitive, carceral approaches tend to exist alongside or in relation to more supportive measures, often facilitated by local government bodies and voluntary organizations, which tend to be motivated by concern for the welfare and well-being of marginalized populations and a recognition that poverty and precarity cannot just be “policed away.”[117]

In addition to illustrating a generalized postrevanchist turn, the involvement of the Larrakia Nation Aboriginal Corporation in the NTASB app and its apparently contradictory framing are indicative of geographically and culturally specific social relations that differentiate not only settler and Indigenous communities, but also particular groups of Aboriginal people from one another. As the NTASB app makes clear, Indigenous peoples living in Australian society constitute a range of populations who are subject to varying degrees of stigmatism, correlating with their participation, or lack thereof, in the mainstream economy and normative (i.e., settler) modes of citizenship.[118] Through surveillance, reporting, and intervention, the app reproduces connections and distinctions between two Indigenous groups in Darwin—“long grass campers and the Larrakia—as represented by the corporation. They are distinguished from one another in terms of their relationship to settler governmental institutions, public discourse, the urban landscape, capital accumulation, property development, and the tourism economy. As Daniel Fisher highlights, Larrakia Nation’s status as intra-Indigenous care givers derives from and affirms the status of the Larrakia as de facto “owners” of Darwin: while a long-standing legal battle between the Larrakia and Northern Territory government saw the former pursue land rights and native titles in the courts, this dynamic has since shifted and, as the NTASB app exemplifies, the Larrakia's status as traditional owners has been affirmed in situations relating to political interests of Northern Territory government, enabling them, via representative institutions, to become "partners" for the purpose of governing a wider Indigenous population.[119] These partnerships are grounded in corporate contracts and privatized relationships rather than a legally granted sovereignty. Yet for Larrakia Nation, such partnerships provide an opportunity to *perform* sovereignty through practices of care. The provision of welfare and pastoral services, such as the night patrol, places Larrakia Nation alongside other governmental forces and institutions. However, as Fisher is keen to emphasize, the "state effect" that arises from Larrakia corporate institutions is not simply driven by an instrumental logic but by an understanding of relatedness between Larrakia and their neighbors, and a respect for campers as fellow "countrymen."[120]

In attaching noise to elements of the population who depart from norms of reproductive respectability and by directing increased policing and surveillance towards them, the NTASB app has much in common with the other manifestations of auditory revanchism discussed in this chapter. Yet the involvement of Larrakia Nation Aboriginal Corporation, and the intra-Indigenous politics of care that accompany the app prevent straightforward comparisons. Instead, the app's conflicted, if not contradictory, combination of criminalization and noncarceral intervention stems from differing settler and Aboriginal interests, and the correlative tensions and compromises that accompany such partnerships.

The "caring" postrevanchism of the NTASB app, however, needs situating within Darwin's neocolonial geographies of policing and punishment. In 2019, the same year the NTASB app was launched, then–NT Chief Minister Michael Gunner sought authorization for private security guards to operate under the Local Government Act, enabling them to deal with breaches of council bylaws in the Central Business District including noise, camping, begging, and loitering in public spaces.[121] While the city's administration has claimed that private security forces cannot enforce bylaws and can only "detect a problem," the Australian investigative journalism platform Four Corners has collated accounts of these forces acting as enforcement, often with recourse to violence.[122] More recently, the age of criminal responsibility was once again lowered to ten years old with the Country Liberal Party's return to government in 2024. Enacted as part of a suite of law and order reforms described as leading to "improved safety, a better lifestyle and stronger economy for all Territorians and visitors,"[123] the move disproportionately affects Indigenous communities in a region that detains children at a higher rate than any other area of the country.[124] In stark contrast to the construct of the (white-racialized) child-as-investment in chapter 3, then, Aboriginal children are construed by the settler state as both vulnerable persons in need of protection from their own families *and* deviant non-children in need of incarceration. Further, a series of high-profile police killings of Aboriginal people in the Northern Territory (including significant numbers of victims with disabilities), has affirmed neocolonial distinctions between protectable and disposable life. Any turn to a "kinder, gentler" (post)revanchism enacted by the NTASB app is necessarily constrained by punitive interventions that seek to "take back" the city from those whose presence degrades it and, more broadly, the endurance of a social and economic compact that demands precarity for some in the name of safety for others.

## SOUNDS OF SURPLUS LIFE

Taken together, the use of classical music as an audio-affective deterrent, the sonic policing of the *pancadão*, and the surveillant listening of the NTASB app constitute expressions of auditory revanchism, whereby sound, music, and aurality become imbricated with a "vengeful" urbanism targeting those charged with "ruining" urban space and threatening financialization's promise of investment, prosperity, and renewal. Auditory revanchism exposes how the drive to "take back" the city (and other urban environments) from unwanted surplus populations also operates on sensory and aesthetic registers. To this end, recorded music, militarized sound, and surveillant listening have been employed against those who breach reproductive respectability and its temporal, spatial, and auditory expressions. These sonic surrogates are not only emblematic of the ways that criminalization has become central to how capitalist societies manage the problem of surplus populations; they also reflect some of the ways that, as a public "service," policing and securitization has itself become increasingly automated, marketized, and technologically mediated.[125] Of significance, too, is auditory revanchism's media culture: the hundreds of news reports on classical music's repellent force, the audiovisual spectacle of *polícia acaba com pancadão*, noise abatement online forums and the social media comments debating the NTASB app imply that revanchist reproductive sound technologies are enmeshed with other modes of mediation that disseminate, intensify—but sometimes critique—their "vengeful" construction of an urban enemy. At the same time, auditory revanchism is often informed by and combined with localized and sometimes contradictory policies, histories, and social relations. As I have repeatedly emphasized, there are substantive differences between Guilliani's zero tolerance policing and New Labour's crusade against antisocial behavior; likewise, there are substantive differences between *Programa de Silencio Urbano* and the noise abatement activities of the Larrakia Nation Aboriginal Corporation. Presenting these exemplars together is not intended to flatten these differences: rather it serves to highlight the continuities and differences in how auditory revanchism occurs across distinctive geopolitical contexts.

While auditory revanchism is partly constituted by vengeful sonic practices of displacement, it also bears the traces of life that persists informally. Since reproductive respectability is defined in accordance with the wage-relation, determining the appropriate times and places for rest and recuperation, and directing appropriate uses of leisure time, it has limited applicability for those who find themselves temporarily or permanently surplus to the requirements of capitalist production—those populations who are unemployed or underemployed and who, via race, class, and coloniality, occupy a position of what Stuart Hall et. al. describe as "secondariness."[126] The sonic and social practices at stake here—loitering, late-night street parties, long grass camping—are expressive of the ways the wageless, while frequently subject to state-permitted violence and government interference, nevertheless foster alternative and potentially frictional modes of spending time and using space. The point here is neither to simply romanticize these as oppositional practices nor disavow the (anti)political agency of urban surplus populations.[127] Rather, my intention is to highlight how, in its encounters with revanchist reproductive sound technologies, the antisocial sociality of surplus life—of populations rendered unproductive, disposable, or expendable, be it by the legacies of colonization, stagnating national productivity, or the rescindment of social welfare—often appears as a noisy remainder, a necessary externality that is itself a structural consequence.

# 5 Lull, Mask, Calm, Renew

## MUSIC AS SLEEP AID

A gently plodding piano melody is joined by regular interventions of a muffled, rumbling bass every six notes on the third beat of the bar. Slow and repetitive, with little dynamic and timbral variation, the opening to Max Richter's *SLEEP* (2015) engineers a sense of stillness in its restraint. This is maintained throughout the eight and a half hours of the piece. Described as a personal lullaby for a frenetic world and a manifesto for a slower pace of existence, Richter suggests that his children were one of the key inspirations for the work: "When I look at my children I wonder where they will find rest. Those moments of being that they used to have as tiny babies, arms outstretched, trusting of the world. I've long wanted to write something that might soothe them."[1] For Richter, the significance of *SLEEP* also lies in its purportedly antagonistic relationship with the conventions of contemporary classical music: "Somehow in Europe, over the last century, as complexity and inaccessibility in music became equated with intelligence and the avant-garde, we lost something along the way. Modernism gave us so many stunning works, but we also lost our lullabies. We lost a shared communion in sound. Audiences have dwindled."[2] *SLEEP* responds to classical music's alienation of its audience with a return to musical simplicity, reuniting music once again with a social purpose.[3]

*SLEEP* now exists across various media. The full version of the composition, which is available as a recording and has been performed live in cities in Europe, the US, and Australia, is intended to be experienced while the listener is asleep. A ninety-minute abridged recording of *SLEEP* caters to those who want to hear the piece while awake and who cannot commit to eight and a half hours of listening time. The documentary film *SLEEP*, released in 2019, features interviews with the composer and follows preparations for and the performance of the piece in Los Angeles' Grand Park. Further attempts to cultivate new uses and audiences for the piece resulted in the release of the *SLEEP* app in 2020 by Richter and Universal Music Group, which was made freely available for those with a Spotify Premium or Apple Music account. Users can select from three activities—sleep, meditate, and focus—each of which activates a specific musical sequence and can be accompanied visually by planetary-themed animations. The app also features a wake-up alarm composed by Richter. If listened to using an Apple Watch, the app uses heart rate data to adapt the music accordingly.

Richter's composition and its associated media commodities draw on the cultural associations between music and sleep, forged by musical styles and practices such as lullabies and nocturnes. Indeed, while it stands out due to its mainstream visibility and popularity, *SLEEP* is one of numerous installations, recordings, and performances intended to aurally induce, shape, and accompany sleep. Among these are Hajime Murooka's 1970s records mentioned in chapter 3, which combine uterine sounds with classical music to encourage sleep in newborns. Although they address different audiences and contexts, there are resonances between their scientistic framing: in the case of Murooka, citations are made to psychology and acoustics, while Richter engages with neuroscience. Richter himself has stated that *SLEEP's* variation form is inspired by Bach's Goldberg Variations and the apocryphal tale of its composition as a sleep aid for the insomniac Russian diplomat, Count Keyserling. More recent examples interrogating connections between sound, aurality, and sleep include Steven Stapleton's twelve-hour sleep concert, a project that began in 2011 and involves Stapleton playing and remixing his own dark, ambient music for a sleeping audience. The soporific potential of musical listening, then, has often been a source of creative inspiration.

Although *SLEEP* is grounded in the long-standing connections between music and rest, its framing and popularity are also expressive of cultural anxieties about the present-day status of sleep. Richter himself has described *SLEEP* as a quiet protest to the hectic pace of contemporary life, where sleep offers a creative space for the mind independent from the demands of wake time: "We have a very screen-based life—we're cogs in a late-capitalist machine, and we're always on. Sleep is an act of resistance against that."[4] Such remarks offer a crude rendition of some of the arguments put forward by critic Johnathan Crary in his diagnosis of "24/7 capitalism." For Crary, sleep offers a significant site of autonomy in an era characterized by sleeplessness. Although he cautions that sleep is by no means an "enclave outside the existing global order,"[5] Crary frames sleep as "unexploitable" and "unassimilable": "Sleep is a remission, a release from the 'constant continuity' of all the threads in which one is enmeshed while waking. . . . Because capitalism cannot limit itself, the notion of preservation or conservation is a systemic impossibility. Against this background, the restorative inertness of sleep counters the deathliness of all the accumulation, financialization, and waste that have devastated anything once held in common."[6]

Richter's framing of his composition makes apparent a particular cultural discourse in which sleep is romanticized: its proximity to the child, for instance, imbues sleep with innocence, simplicity, and freedom, while sleep's association with dreaming enables its framing as a site of creativity. The subject who sleeps is unmarked: a general victim of (late) capitalism and technology, a shared "we." Yet the presentation of sleep as both a universal problem (due to the pace and demands of capitalist life and its technological mediations) and mode of respite (where we can slow down, dream, have creative ideas) omits that, for many, it is not screens but material precarity that constricts sleep's promise of renewal. Indeed, the tension between Richter's idealized framing of sleep and the realities of sleep for many poor, urban populations are briefly gestured to in August Brown's review of *SLEEP*'s performance in Los Angeles' Grand Park: Brown reminds the reader that "almost 58,000 homeless people are estimated to bed down on the streets of Los Angeles County on any given night," adding that he hopes "this fact is not lost on the 500-plus people who on Friday night took to cots in downtown's Grand Park."[7]

Sleep has been a recurring yet largely implicit theme of this book's previous chapters, connecting the robot's mechanical lullaby in chapter 1, the Foxconn factory workers residing in dormitories and the replacement of night-shift carers by listening devices in chapter 2, the restful newborns of chapter 3, the disrupted nights of São Paulo's periphery and the "antisocial" rest of long grass campers in chapter 4. Sleep is, after all, integral to processes and practices of reproduction: periods of rest are required for the worker to work, for the child to develop, and for the population to remain mentally and physically well. As exemplified in chapter 4, the space and time of sleep is bound up with exertions of reproductive respectability: punitive measure are taken against those who do not use the designated time of sleep for that purpose, those who disrupt the sleep of others, or those whose sleep takes place outside and in public. Yet the boundaries between sleep and wakefulness, work and rest, day and night have often been porous, historically contingent, and subject to transformation. As Julia Morandeira makes clear, preindustrial or feudalist nighttime was by no means exclusively dedicated to sleep: times of sleeplessness could be used for prayer, domestic labor, intimacy, thought, or petty criminality.[8] The transition, which E. P. Thompson describes, from task-oriented time to clock time, and with it, the establishment of industrial capitalism's puritanical time-discipline necessitated by the need to synchronize labor, served to standardize a dialectic of day-time work and nighttime rest.[9]

At the same time, the length of the capitalist working day has undergone periods of expansion and contraction, often through a combination of class struggle and new technologies: innovations in artificial lighting, for example, meant that the rhythms of labor could be separated from diurnal cycles and the working period could be extended into the night.[10] The globalization of the workforce, similarly, enables companies to work around the clock, with one set of staff ending their working day as another begins. And if the time of production has been extended and expanded, so too has the time of consumption, with, for example, the formalization of nighttime economies and sleepless cities often as an accompaniment to urban renewal (see chapter 4).[11] Yet if, as Silvia Federici asserts, the body is "the condition of the existence of labour-power" and "also its limit," then sleep remains one of the motors of capitalist society, even as its temporal designations have become destabilized by both the globalized expansion of the working day and the round-the-clock consumption offered by the "twenty-four-hour city."

The accompanying pamphlet to the recorded version of Richter's *SLEEP* opens with an epigraph attributed to Heraclitus: "Even a soul submerged in sleep is hard at work and helps make something of the world." I propose that the transformation of *SLEEP* into an app helps ensure that the "soul at sleep is hard at work." To this end, I want to suggest that, contra claims of sleep's distance from "accumulation, financialization and waste," Richter's app is expressive of wider trends within a long-established sleep industry. Ultimately, such commodities aim to smooth out the contradiction between the demands of capital and the requirements of the body by making the latter a driver of accumulation and consumption. Sound and music have had a key role within these efforts, reflected in the innumerable sound-based sleep aids now available on app stores, streaming services and online shopping sites, and I contend that their promises to soothe, relax, and lull are grounded in sleep's historical connections to gendered divisions of affective and musical labor. While many of these devices incorporate low-cost and royalty-free sound recordings, some recent sound-based sleep aids have involved more direct interventions into the production of music. To this end, I offer an analysis of the streaming service app Calm, and the convergence of the recording and digital "health and wellness" industries. Beginning with the lullaby and ending with its automated "app-lication," this chapter thus sounds out the place of sleep and its musical accompaniments within shifting formations of production and reproduction.

## LULL: MARITAL SECRETS AND MUSICALIZED CARE

Anyone who has been involved in caring for one knows that sleep rarely occurs by simply leaving a newborn to their own devices. Although it often appears as a natural, biological instinct, sleep is a practice that is cultivated. Much of the infant's circadian rhythm, which helps consolidate a diurnal sleep-wake cycle, develops in the first three to four months of life, arising from a combination of physical factors (e.g., cortisol and melatonin fluctuations) and environmental signals (changes in light and sound levels).[12] During and after the establishment of the circadian rhythm, caregivers may try to guide infants in distinguishing the time for sleep, and adopt cues and routines that help young children to drift off and transition from wake-time.

In many cultures, lullabies are an important resource for caregivers wanting to cultivate sleep. Simultaneously signifying, somatic, and affective, and often categorized as a folksong, the lullaby marks out the time for sleep by creating a soporific atmosphere. It is typically a quiet, slow, and repetitive song intended to soothe, calm, or lull the listener, and is often accompanied by rocking or another form of rhythmic movement. It tends to feature hushed, melodic vocals and a repetitive form; however, the lullaby's lyrical and thematic content can vary widely. Lullabies have a long history: while no compositions survive from the time, there are references to the use of such songs in domestic settings and within the family sphere that can be traced back to Greek antiquity.[13] They are also geographically widespread: gentle songs for inducing sleep can be found across the world. However, lullabies are also culturally specific practices, with lyrics often incorporating references to local folklore, identities, and knowledge. Furthermore, lullabies frequently draw upon sleep's proximity to death and darkness. There are thematic connections to both murder ballads and laments, with stories of demise and expressions of sadness having a strong presence within the repertoire. The Japanese lullaby "*Takeda no komoriuta*," a recording of which became popular in Japan in the 1970s, is a nursemaid's lament for home: the song expresses her unhappiness as the child often cries, and she longs to go home to her family. Threats of being harmed by animals and spirts should the listener not go to sleep are similarly common. The Haitian creole lullaby "*Dodo, ti petit manman*" (Sleep, mummy's little one) or "*Dodo titit, krab nan kalalou*" (Sleep my little one, crab in callaloo) is instructive in this regard: it contains the sweetly sung line: "*si ou pa dodo, krab la va manje ou*" (if you do not sleep, the crab will eat you).[14] Some lullabies even express violent wishes towards a (non-sleeping) infant. A Scandinavian lullaby, for instance, contains a line that translates as "If the child won't sleep grab it by the leg and bang it against the wall and then the child will sleep."[15]

Lullabies are typically performed by and associated with feminized caregivers: most prominently mothers, but also grandmothers, sisters, nurses, maids, and nannies. This gendered designation is by no means universal.[16] Some traditional Irish lullabies feature an ambiguously gendered or male narrator: the song "Rocking the baby that nobody owns," for example, both maintains and subverts the common gendered associations of the lullaby, with a husband caring for a child that is not his.[17] Narrated by "an old man," the song is a cautionary tale about infidelity: his wife stays out all night, while he is left by himself, rocking the cradle of a baby that he has not fathered. "Rocking the baby," then, is an expression of masculine resentment towards "bad" wives and mothers who fail to fulfill their marital and domestic duties, among which, it might be assumed, is the singing of lullabies for their children. Indeed, their connection to caregiving, childrearing, nursing, and other types of reproductive activity, means that the performance of lullabies is determined by gendered divisions of labor. It may therefore be categorized as a type of work song, as Holly Pester proposes: "Like any shanty or marching chant the rhythms of her body and the tempo of the song—rocking and jigging the baby into slumber—co-ordinate the act of material effort (in the scene of supposedly immaterial labour)."[18] This act of material effort—a mode of musical labor—produces the sleeping baby: the exertion of one produces rest for another. And in generating sleep, the lullaby ensures other work can be done, either by providing time for other tasks to take place (as articulated in "dodo titit," in which the infant needs to sleep as mother is away at the market and father is away at the river) or alternatively, in allowing other household members to also rest.

"Rocking the baby that nobody owns" is illustrative of another significant tendency of lullabies: its use as a means of disclosing secrets to an infant who can listen but cannot understand nor repeat the information they are being told. Although she is never referred to as such (only as "my wife"), the narrator tells the child of their mother's adultery, while also speculating as to whether they will ever meet their biological father. The treatment of the lullaby as a means of disclosing secrets arises largely from its performance context: they are typically sung when infant and caregiver are alone, together. Thus, as Barbara Hillers argues: "The singer is free to speak her mind; even if someone overhears her, her performance is not likely to cause offence, since she is after all doing her duty in soothing the baby."[19] In such instances, the infant listener becomes an affective vessel, while their assumed capacity to hear without speaking makes them an ideal confidant. The intimacy of the lullaby, then, arises not only in its musicalized expression of care but also in its capacity as a medium of disclosure.

In articulating a caregiver's feelings of love and fear, betrayal and frustration, and trying to encourage sleep through reassurance or threat, the lullaby is characterized by various disjunctures. There is an apparent disjuncture between the lullaby's affective function as a soothing, lulling song conducive to rest and some of its common lyrical sentiments, which tell of threats, tragedies, and unhappiness. There is also a disjuncture between the lullaby's appearance in the popular imagination as an auditory expression of motherly love or an idyllic expression of the maternal-infant bond and the wide-ranging and often far-from-idyllic sentiments and stories that are contained within the genre. This disjuncture between the idyllic lullaby and the lullaby's varied sentiments shadows the disjuncture between an idyllic maternalism and the lived realities of caregiving. Lullabies express the isolation, loneliness, exhaustion, and worry that are felt by those who act as caregivers to children, sentiments that are often attached to experiences of poverty and other types of social precarity. These negative and ambivalent sentiments are obscured by idealized constructions of motherhood and maternalism, which center self-sacrifice, enjoyment and satisfaction in relation to the nuclear family. As Barbara Hillier notes, "the lullaby is yet another genre of oral tradition which has been sanitized for consumption in the bourgeois nursery,"[20] resulting in the concealment of its unsentimental, taboo, or discordant elements.

The sanitation of the lullaby is recognizable from recent initiatives that mobilize lullabies in health and wellness contexts, emphasizing their soothing capacity and its ability to enforce bonds between (maternal-familial) caregiver and infant. Such is the case in the Lullaby Project, a participatory arts initiative that partners new and expectant parents who are vulnerable or minoritized due to their socioeconomic circumstances with professional musicians to compose a personal lullaby for their child. The program began at Carnegie Hall's Weill Music Institute, where its aims were described as supporting maternal health, aiding childhood development, and strengthening the emotional bonds between parent and child. A similar—and revealing—description of the project was offered in the

*New York Daily Post*: it helped, in the words of a Department of Corrections official, "women who have made some bad choices" not only to "be good parents" but also to "strive for better futures," both for themselves and for their offspring.[21] A report detailing the Lullaby Project's impact and outcomes credits the project with the development of a positive mindset among participating mothers: outside of the project, participants spoke of the social and economic hardships they faced and the challenges, fears, and anxieties of parenthood, whereas within "the lullaby setting" participants shared "positive emotions and relationships, a new or re-discovered sense of personal meaning and achievement, new insights, and hopes for the future. These are all signs of well-being."[22] The project has since been taken up in the UK, engaging prisoners, perinatal mental health service users and refugees through partnerships with the Irene Taylor Trust (a charitable organization focused on music in prisons), Live Music Now (a charity that works with those who have reduced access to live music), RPO Resound (a community and education program of the Royal Philharmonic Orchestra), and the National Health Service.

The Lullaby Project reinvents the lullaby as a personalized and predominantly aspirational medium that expresses parental hopes and dreams and, in so doing, serves to strengthen familial bonds. As the remarks of the Department of Corrections official suggest, its function is not so much the production of sleeping infants but transforming "vulnerable" or "at-risk" mothers into "good" parents. In describing the project in such terms, I do not intend to discount the value of providing new and expectant parents experiencing social isolation with creative opportunities and spaces of self-expression. Nonetheless, the project is imbued with a particular classed, gendered, and racialized ideal of motherhood.

This motherhood ideal is traceable in a 2017 report into a UK pilot of the Lullaby Project, which refers to the significance of secure attachment. The report defines attachment as "a deep and enduring emotional bond that is known to be of central importance for healthy psychological development. Secure attachment styles are promoted in part by effective early bonding between mother [*sic*] and child, implying adequate responses to the child's expressed needs."[23] Such uses of the term broadly align with the central premises of attachment theory. Originating in the post–World War II research of British psychologist John Bowlby, attachment theory concerned the infant's apparent drive to secure maternal response through a set of evolutionary behaviors originating at birth, motivated by fear, affection, exploration, and caregiving.[24] Bowlby states:

> Essential for mental health is that the young child and infant should experience a warm, intimate, and continuous relationship with his mother (or mother-substitute) in which both find satisfaction and enjoyment. Given this relationship, the emotions of anxiety and guilt, which characterize mental ill health, will develop in a moderate and organized way.[25]

Bowlby's theory of attachment emerged a time when questions about mothering and women's work roles were being widely debated, contributing to an intensifying discourse of maternal blame. And while Bolby is clear that a "mother-substitute" could undertake the maternal role vis-à-vis attachment, as Marga Vicedo suggests, attachment theory's naturalization of maternal love and care nevertheless offered a renewed justification for gendered parental roles predicated on ideals of white, middle-class subjectivity.[26] In crediting the maternal-infant dyad as the source of psychological wellness for both caregivers and their children, then, the Lullaby Project reproduces some of the central precepts of attachment theory.

The Lullaby Project attests to contemporary efforts to expand and capitalize on music's uses as a tool of well-being: a broadly defined indicator that, through its commodification, has come to signify a range of affective states to be promoted (e.g., resilience, positivity, relaxation, fulfilment) and avoided (e.g., stress, depression, mental ill health), explicitly or implicitly in relation to various social outcomes (e.g., the ability to work, undertake care work, remain in education, live "independently"). As will become apparent over the course of this chapter, the production and management of well-being—via music and other means—is now a key rationale for many digital products and services. Music's deployment as a tool of well-being is also traceable within charitable and "third sector" settings, workplaces, and in health and social care settings (e.g., see chapter 3). While it is sometimes an analytic of the collective, such interventions typically pertain to the individual: well-being is something that one possesses and, importantly, is responsible for. Music's ability to bolster well-being has become particularly prescient as state institutions of social reproduction (health care, social services and so on) are increasingly stretched, and the Lullaby Project's recent engagement with English perinatal mental health service users is instructive in this regard.[27] However, the pursuit of well-being on behalf of particular social and medical communities can open up new partnerships and "diverse" revenue sources and income streams, and to this end, it has also become prescient as music itself faces its own struggles for economic security: although they have largely remained successful in generating investment, institutions like Carnegie Hall and the Royal Philharmonic Orchestra face a challenging landscape in which public and philanthropic funding is highly competitive. It is within this context that musical "outreach" initiatives such as the Lullaby Project need to be understood.[28]

The Lullaby Project is one of various efforts to put lullabies to work in contemporary lives. However, whereas the Lullaby Project ultimately treats music as a charitable gift (the lullaby is exchanged for a participants' time and the opportunities for evaluative reporting that follow from it), lullabies frequently circulate either as, or as part of, a commodified good or service. As noted in chapter 1, the production of lullabies for children was one of the (many) uses proposed by Edison for the early phonograph and, sure enough, lullabies are now often produced and disseminated as musical recordings. They are also prevalent on streaming services and within app marketplaces, where collections of recorded lullabies and popular songs in the style of lullabies appear alongside the sounds of air conditioning units and rainforest soundscapes, intended for both children and adults alike. By simply switching on (tapping a screen or setting a timer), the user can switch off.

## MASK: SOUND AND THE SLEEP INDUSTRY

If there is nothing new in having trouble sleeping, then there is also nothing new in seeking auditory solutions to such problems. Even those in close contact with the divine have turned to sound when challenged by the inability to sleep. In 1667, the Italian sculptor and architect Gian Lorenzo Bernini was called upon by Pope Clement IX on the day of his election to create a tool to help him to sleep. According to Bernini's son, the pope had already tried procuring sleep with the help of external means, by sleeping in bedrooms that were in close proximity to running water. When he was raised to the papacy, the pope asked Bernini to redirect "all the water that flowed in the various fountains in the Belvedere garden to a single fountain that stood immediately below the window of his bedroom so that he could better hear the sonorous noise of the cascading waters and thus more easily achieve a state of repose."[29] As this request was not possible to fulfill, Bernini instead created a machine that imitated the sound of flowing water. The device featured a wheel in combination with a series of paper globes that produced a sound when struck, and was placed in the room adjacent to the papal sleeping quarters.[30] The machine is said to have worked, with the pope lauding the genius of Bernini, which "expressed itself in the same way in little things as in great ones."[31]

The pope's preference for the sounds of running water, and by extension, the efficacy of Bernini's machine, likely relate to their ability to block out, mask, and thus ameliorate sleep-preventing disruptions. It is this masking principle that underpins more recent sound-based sleep aids, whereby recorded sound and music are used to conceal auditory intrusions from neighbors, passing vehicles, humming boilers or the twenty-four-hour city. Equally, sound-based sleep aids can also conceal an uncomfortable silence that, during the dark hours of the night, can accompany a racing mind or amplify the internal sounds of tinnitus. To note that there are functional similarities between Bernini's device and modern sound-based sleep aids, however, is not to downplay the important social transformations that prevent us from making simplistic conflations. Among these transformations is the medicalization of sleep and sleeplessness, with which sleep becomes a pathological "problem" that needs rectifying.

In a lecture printed in the esteemed medical journal, *The Lancet,* Sir William Broadbent opens by stating, "Sleeplessness is one of the torments of our age and generation."[32] The sentiment behind this proclamation is likely familiar to many twenty-first century readers. Yet Broadbent's "age" was the start of the twentieth century: the lecture was given in 1900, by which point sleep and sleeplessness had become established areas of clinical interest. While the English term "sleepless" originates in the Middle Ages, the notion of insomnia, or sleeplessness as a state of being, are largely products of the Enlightenment, which is to say an increasingly rationalistic culture that understand problems in scientific terms.[33] The insomniac, meanwhile, is primarily an invention of late-nineteenth century North American and European culture, in which sleeplessness became a disease in its own right, as opposed to a symptom of an underlying medical, psychological, or social condition.[34]

For Matthew Beaumont, the rise of the insomniac as a social archetype marks the origins of the current sleep crisis, "in which anxieties about sleeping are inextricably tied to concerns about new technologies and the accelerating velocities of everyday life can be located."[35] The insomniac was the result of the rush and stimulation of modern life and closely associated with states of nervousness, neurosis, and intellectual overwork. It was—and to some extent remains—a masculinist and middle-class disposition. As Beaumont argues, the poor could not afford insomnia, "for however inadequately they sleep during the night it is pointless to complain because they are under compulsion to rise again at a decent hour the following day and resume their physical labour."[36] Instead, sleeplessness was connected to the overworked mind of professional men: a culture of long working hours combined with little sleep could lead to a diminishment of mental abilities.[37] Scientific interest in women who experienced sleeplessness was relatively rare, not least because, as the history of the lullaby attests, sleepless nights are simply part of the job of caring for restless infants or soothing sick children.

If the modern crisis of sleeplessness is largely a late nineteenth-century invention, then so too, is the sleep industry that promises to ameliorate its effects. As Bernini's device exemplifies, the history of sleep aids spans many centuries, with evidence of sleep medicines dating back to at least the sixteenth century.[38] However, the rise of sleeplessness as a predicament of modernity was accompanied by an expanding field of doctors, inventors, journalists, and cultural commentators offering solutions to those needing help in achieving nighttime rest. These solutions included both pharmalogical and nonpharmalogical interventions, such as diets, rituals (such as reciting the Lord's Prayer with eyes downturned), breathing exercises, and particular sleeping arrangements (such as assuring one rests on a cool pillow).[39] In addition, sleeplessness became one of the maladies targeted by electrotherapeutic devices such as the Harness Electropathic Belt: a late nineteenth-century British product made by the Medical Battery Company that was marketed as a nonpharmacological device capable of imparting "new life and vigour" by ameliorating "nervous exhaustion, neuralgia, rheumatism, indigestion, sleeplessness, ladies ailments" and more.[40]

Patents for sound-emitting devices that help induce sleep date back to at least 1894.[41] However, the incorporation and expansion of sound technologies within the sleep industry primarily occurred later in the twentieth century. James K. Buckwalter has been credited as the inventor of the first domestic white noise machine. Purportedly inspired by the absent sound of a broken air conditioning unit, Buckwalter developed a prototype circa 1960 constructed from a fan blade, a turntable, a dog bowl and some insulating foam, submitting a patent for the "sound-producing sleep-inducing device" in 1963. The originary status often attributed to Buckwalter's invention is troubled by Bernini's much earlier machine, and the patent for the "sound-producing sleeping-inducing device" makes clear there were other conceptual predecessors from the postwar era: these include Charles Beazley's 1950 patent for an "acoustic sleep-inducing apparatus" and Frank H. Weisman's 1952 patent "device for inducing sleep and rest," which is described as suitable for use in hospitals, as well as in the home.[42] Nonetheless, Buckwalter's became one of the first mass-produced and commercially successful auditory sleep aids. By the second half of the 1960s, Marpac's "Sleep-Mate" was being marketed in domestic appliance catalogues aimed at US middle-class household.

Whereas the late nineteenth-century insomniac was predominantly a masculinist archetype, whose trouble sleeping was an undesirable consequence of their pursuit of intellectual labor, the Sleep-Mate was constructed as an explicitly feminized technology. The device's marketing materials routinely show white women sleeping peacefully next to the siren-shaped sound conditioner, their smiling faces turned towards the machine. It is depicted as an unobtrusive presence in the intimate space of the bedroom, where the masking capacity of its mechanical noise serves to generate the acoustic boundaries necessary for suburban quietude. Outside the bedroom, it is suggested that other household members could benefit from the Sleep-Mate's ability to mask auditory intrusions through its "soothing sound of rushing air." A Sears Roebuck 1966/1967 catalogue advert shows both a sleeping woman and a child sitting at a desk reading: the accompanying text suggests the device "may help you relax or sleep and even calm down baby. Study with Sleep-Mate . . . it muffles sounds that break concentration."[43] The apparent contradiction in the Sleep-Mate's proposed domestic uses—its ability, through auditory masking, to facilitate rest *and* concentration—prefigures many contemporary devices: as Richter's app version of *SLEEP* attests, the combination of "sleep" and "focus" (along with "relax" or "meditate") frequently appear as target-activity states to be facilitated by sound-based sleep aids, intimating that the crisis of sleep is also a crisis of distraction. Avoiding possible confusion about the purpose of its auditory masking technology while also expanding its range of products and potential consumers, Marpac ended up essentially producing the same device for different contexts. The Sleep-Mate was marketed and framed as a technology for the home, while the "Sound Screen" was framed as a study aid that could be used in educational and professional settings.

In his historical investigation of the Sleep-Mate, Mack Hagood describes how the device circulated as a health and well-being product that offers self-regulation and control. For Hagood, the Sleep-Mate can be recognized as a "spatial technology of the self" that produces privacy by impairing audition, and thus offers freedom *from* hearing. Drawing on Buckwalter's original patent, Hagood highlights how the Sleep-Mate relies on the rhetorical invention of a self that needs such a device in order to achieve "natural sleep." Buckwalter's construction of the self is, in Hagood's words,

> thoroughly modern, containing the contradictions of a liberal, capitalist society. This self is highly autonomous, responsible both for its own "self-generating" distress and for finding a patented technological solution. Yet for all its autonomy, this self is also highly sensitive to the strangeness of its environment, requiring "exclusion" of what the patent calls "sound of disturbing character" (one imagines it lurking outside in a trench coat) in order to achieve sleep. Tellingly, this exclusion is achieved through the manipulation of sonic space because the self is not able to regulate its own boundaries alone. Herein lies the modern contradiction: the modern self requires autonomy yet is too porous and inhabits a space too penetrated by stimuli, to achieve autonomy on its own.[44]

Hagood is correct in his observation that a particular figuration of the self is central to the Sleep-Mate's formation and circulation as a media technology. Yet an exclusive focus on the promise of self-control risks obscuring the significance of other imagined listeners that appear in the device's advertising discourse: namely, infants and school children, both of whom fall under the remit of the suburban, postwar wife. As was made apparent in chapter 1, the family, and the mother-child relationship specifically, has been central to the marketing of household sound technologies since the early twentieth century, helping to make them legible as domestic tools. To this end, the Sleep-Mate's grouping of the sleeping self, soothed infant, and focused child is exemplary of the recurring attribute of reproductive sound technologies, whereby the family form and childrearing activities feature in the design and circulation of auditory technologies that ostensibly offer the individual listener opportunities for self-regulation. Indeed, as I proposed in chapter 1, these domestic figures—mother, child, infant—trouble the conflation of user, listener, and self that frequently grounds cultural analyses of consumerist sound technologies: a conflation that is simultaneously produced and undermined by the Sleep-Mate.[45]

Marpac's Sleep-Mate now exists as Yogasleep's Dohm sound machine. The device's design remains similar to the original, and it still generates noise by electromechanical means. Today, however, the Dohm now faces competition from a plethora of sound-based sleep aids produced by consumer electronic companies, tech startups, software developers, and baby care businesses. Unlike Bernini's and Buckwalter's sound-producing inventions, the vast majority of these sleep aids are based around sound reproduction and the playback of audio recordings. With some exceptions (and discounting the sleep playlists on platforms such as Spotify and YouTube) the contemporary sound-based sleep-aid market primarily consists of custom speaker systems (i.e., white noise or sound machines, specialist headphones, and earbuds) and apps for smartphones, tablets, or smart speakers. Despite the sheer quantity of products now available, there is limited variation in their form and function, with apps and speaker systems typically offering users access to a sound library that allows them to select a preferred audio accompaniment from a range of options.

In keeping with both the twentieth-century Sleep-Mate and Bernini's seventeenth-century device, the promise of a good night's sleep is often attached to broadband noise. The sound libraries of sound-based sleep aids allow users to select from various categories of broadband noise, often identified in accordance with color (i.e., white, pink, or brown noise) or sound source. Common options include the sounds of fans and air conditioning units; "nature" sounds such as waterfalls, wind, and rainfall; or sounds associated with the voice such as breathing and shushing.

There are other types of audio that commonly feature among contemporary sleep aids. These include music. Some sound-based sleep aids offer lullabies, although these tend to be devices targeted at infants (or infants as well as adults), and these lullabies are typically instrumental rather than sung.[46] In devices targeted primarily at adult listeners, the lullaby is often replaced by the more general rubric of "relaxing music," typically consisting of royalty-free library music. This "relaxing music" departs from the lullaby's sung form, substituting anonymized music production for the singing feminized caregiver, but maintains some of its other formal qualities: it tends to be repetitive, slow-paced and gentle, with little dynamic, timbral, and rhythmic variation. Unpitched percussion is rare, while strings, pianos, and synth pads are common. The inclusion of vocals is infrequent and sung lyrics rarer still. In some instances, a crude orientalist aesthetic is at play, with the apps' sonic materials and playback categories reproducing the pernicious conflation of "exotic East" with relaxation.[47] Listeners can often select a custom length of time—from a few minutes to hours—so the sleep aid's relaxing music is often composed to be looped. Just as the lullaby can be sung over again until the infant is asleep, the sleep aid's relaxing music can be repeated as many times as needed.

Other common types of sound featured in contemporary sleep aids include "soundscapes" (a term that is used in this context to refer to a sequence of sounds associated with a particular environment such as a city or a forest), binaural beats (an auditory illusion produced by listening to two different frequencies, which has come to be associated in the popular imagination with brain wave entrainment), and spoken word recordings that take the form of audiobooks, guided meditations, or relaxation exercises. Morphée, a sound machine produced by a UK-based company of the same name is illustrative of the latter. Notably, the company differentiates its product to prospective purchasers by touting the fact that it does not require them to use their phone as they are trying to fall asleep: rather, the device is described as "non-digital, wave free and without screen for optimal efficiency."[48] As such descriptions suggest, Morphée attempts to present their device as a comparatively organic, or "non-technological" technology. Appearing as a compact music box with three turning keys, the devices' sound library includes guided "visualizations," breathing exercises, and meditations. Users are able to select between a "male" and "female" voice, as well as use the device to listen to "relaxing music" or "nature sounds." The company also offers "my little Morphée": an alternative device with which users (i.e., caregivers or children themselves) can select an "animal companion" to guide the listener through various "meditation journeys" and which is intended to "help . . . children calm down and fall asleep."[49]

Sound-based sleep aids might be categorized as a type of ambient media, intended to modify the auditory environment and cultivate a type of listening that is intentionally inattentive.[50] Their sounds are not primarily intended to be listened to but rather, serve as a background, either concealing other disruptions or masking an uncomfortable silence. This characterization, however, is troubled by Morphée's voice-based device. In this instance, the listener is to focus on the voice and follow its instruction. Listening in this context begins as attentive, although the aim is to progress to a state where this is no longer the case, with the listener falling asleep. In her 2013 assessment of sound and music apps for sleeping, Anahid Kassabian notes that the voice is both prevalent among sleep apps and makes it challenging to generalize about their form and function. However, she maintains that voice-based devices share with other sound-based apps the notion that certain types of sound are conducive to rest. To this I would add that Morphée and similar devices make apparent that sleep aids may both operate in relation to multiple types of listener (i.e., children, babies, adults) and seek to cultivate multiple types of listening practice, occurring along—and sometimes moving across—the spectrum of foreground-background, attentive-inattentive, or conscious-unconscious.[51]

Where some sleep aids have been framed as allowing users choice and customization—by, for example, allowing listeners to select from various types of sound or even combine or "mix" different sound sources—others have been marketed in terms of adaptability. Kokoon, a technology company founded in 2013, has produced a series of "sleep sensing" headphones to be used in combination with the Kokoon app that allow the user to "master [their] sleep." In 2023, and after a successful crowdfunding campaign through Kickstarter, the company partnered with the conglomerate Phillips to produce their latest model of sleep headphones. Listed as costing £249/$249 when the product first came to market, the Phillips-Kokoon headphones are a high-end product, seemingly targeted at adult professionals (presumably without caring responsibilities for others who might need to be heard in the night). These headphones feature an in-ear heart rate sensor, while the app uses data from this sensor to infer when the user enters different sleep stages. This data offers users the opportunity to track and monitor their own sleep using the "AI-powered" Kokoon app, while also enabling Kokoon's "adaptive audio" function. In practice, "adaptive" in this context means initiating a fade: as the device registers signs of the listener falling asleep, it fades out the audio selection and replaces it with masking noise.

If sleep has been associated with women, inasmuch as the provision of an auditory accompaniment for sleep has historically been a feminized task, then this gendered history seemingly finds new articulation the visual aesthetics that accompany many sleep aids. Women are frequently presented as the primary benefactors of the sleep aid: as with Marpac's Sleep-Mate, sleeping women feature heavily in the marketing materials of various devices. In the instances that women are awake, they are typically accompanied by peacefully sleeping infant or smiling children. There are, of course, exceptions. A 2023 advert for Kokoon's Kickstarter campaign features a middle-aged man tossing and turning in bed next to a peacefully sleeping woman. Kokoon's founder "Tim" intervenes: clad in a white lab coat, he hands the man a pair of headphones, and thus resolves his agitated sleeplessness.[52] Tim's white coat marks the significance of a presentational scientism for many contemporary sleep aids: Kokoon describes its product as "backed by science," while Morphée describes its products as informed by the research of sleep scientists and sophrologists. Consequently, sound-based sleep aids intensify the sleep industry's long-standing, antagonistic construction of sleep as a natural and universal desire and as a practice that can be improved through the application of expertise and commodified intervention. And while for many of these interventions, the features of the audio provided is secondary to the means of mediation (e.g., specialist headphones, "mechanical" music boxes) or the desired function (e.g., the production of relaxation, masking unpredictable sonic intrusions), some recent digital services more clearly center the "sound" of sound-based sleep aids, providing expansive audio libraries and "exclusive" music to users. To this end, the sound-based sleep aid has also become a product of the music industry: a convergence to which I now turn my attention.

## CALM: MUSIC'S APP-LICATION

In March 2021, Calm, "the number one app for meditation and sleep," and Universal Music Group (UMG) published a press statement announcing a partnership resulting in a new "Sleep Remix Series." Consisting of seven sixty-minute remixes of songs by Universal artists including Ariana Grande, Katie Perry, Jhené Aiko, and Post Malone and launching on World Sleep Day, the series marked the app's growing role within the music industry. Since 2019, Calm has exclusively released a series of tracks and albums on its platform by high-profile artists. This includes recompositions of popular songs by Sam Smith and Sabrina Carpenter in a lullaby style; albums by Moby, Sigur Rós, and Alanis Morrissette; and a new "sleep song" by country singer Keith Urban.

Calm's founders have described its expansion into music provision as a strategic move that helps make taking care of one's own mental wellbeing accessible to a wider audience. Music's ability to modulate emotional states and its popular appeal mean that it is envisaged, in the words of Calm's Head of Music, Courtney Phillips, as "a tool to help people live their lives in a healthier, more productive way."[53] Yet the app's move into music also offers new creative opportunities for artists who want to move beyond the formal constraints of the three-to-four minute song format, historically commanded by radio and inherited by streaming services.[54] In an interview with *Rolling Stone*, Calm's founder and co-CEO Michael Acton Smith suggests the efficacy of the "Sleep Remix Series can be traced to the music's structural elements. Sleep playlists on other apps and platforms tend to be made up of shorter songs or looped audio. These, according to Acton Smith, can prevent the listener from entering what he describes as a "flow state," claiming that a person's brainwave activity appears more chaotic when listening to a compilation of shorter songs.[55] By contrast, Calm's remit allows musicians and producers to create longer musical pieces. However, like the three-minute pop song, Calm still intends its musical releases to gain attention. Calm's music for sleep needs to hook listeners, so that they play the songs in the first place. By grabbing listeners' attention, music can help Calm's subscribers to "stop thinking about [their] to-do lists and that silly comment . . . made at work."[56]

Founded in 2012 by Acton Smith and Alex Tew, Calm is described as a mental health app that provides mindfulness content that can help users manage stress, sleep better, and live happier and healthier lives.[57] Like many other streaming services, Calm offers both a paid subscription and "freemium" model with restricted content. Its users are described as ranging from "little kids" to "stressed executives"; accordingly, the app offers a family subscription model and Calm Kids content. The company has also sought to cultivate markets beyond the private household: to this end, there is Calm Business, allowing companies to subscribe to the service on behalf of their employees, providing access to both health and wellness resources and behavioral and engagement analytics; and Calm Health, a digital mental health screening service for employee-benefit and health insurance programs, which is described as offering "tools and resources to support members during the time between recognizing a need for and accessing professional care."[58] Investors have suggested that Calm has been profitable since 2016, and in 2022 was reportedly valued at $2 billion, with 4 million paid subscribers to the app (the company also laid off 20 percent of its workforce in 2022, citing the "economic climate").[59] Its success has been attributed to the widespread mental health "crisis" stemming from the disjuncture between rising rates of mental ill health and the limited availability of public and community service provision, exacerbated by the COVID-19 pandemic.[60] Initially intended as a meditation app, Calm's function apparently pivoted when the company noticed an uptick in user activity at night implying that listeners were using the voice of meditation instructors to fall asleep, resulting in the platform developing "sleep stories."[61] Since then, its audio content has expanded to include music, soundscapes and voice-guided movement routines for various activities and contexts. There is, for example, "Mindful Running Music" playlist, which promises to help "bring intention to your running routine with energizing beats and calming rhythms designed to unite your breath, body and mind." Alternatively, in the audio library's "For Work" section, there are "Quick Breaks" and "Get Focused" meditations, as well as a "Walk for Mental Health" playlist, featuring a combination of "the sounds of New Zealand and lofi created by Calm's Generative Music Project."

Although Calm suggests that the app provides audio content for various situations and listeners, the aesthetic homogeneity of its music provision is striking, particularly given the vast amount of audio available to users. As with other sleep aids, the music tends to be slow-paced, with few dramatic timbral or dynamic shifts, and piano, synths, and strings are common, as are "nature sounds" (there is a lot of birdsong). Particularly noticeable are the shared production conventions. Take the sleep remix of Post Malone's "Circles," which is just over an hour long. The first five minutes of the mix remains similar to the original song, including the vocals and beat, and then the drumbeat disappears and the sustained synth sounds become more prominent in the mix, and the vocals and the central guitar riff are equalized to become more muffled and muted. By the thirty-minute mark, the song has become barely recognizable, consisting of a gentle, repetitive piano melody and distant-sounding ambient synth tones. The track finishes with drawn out and slow-moving tones that are heavily processed with reverb and delay. These effects are frequently heard throughout Calm's library.

As I noted before, many sound-based sleep aids make use of royalty-free library music by anonymized producers. Calm's music library, by comparison, largely consists of tracks by named artists, with access "rented" via a user subscription. In this regard, Calm is more akin to a music streaming service, although it is not clear what type—or types—of payment model the app uses: for example, whether the artists are offered a flat fee in exchange for exclusive licensing or whether, in keeping with other streaming services, artists receive a fraction of payments made to rights holders based on the proportion of total streams achieved by their sound files. Universal Music Group (UMG) has described Calm's Sleep song remixes as "a wonderful example of how we are working with our artists to create new commercial opportunities by reimagining music."[62] But neither UMG nor Calm have offered information on the terms of their partnership, including how those making music are remunerated for their work.[63]

Calm's partnership with UMG once again illustrates the ways that contemporary consumer technologies and media discourses place emphasis on music's functionality, treating it as a cheap yet sufficiently intimate solution to problems pertaining to social reproduction: in this instance, the apparent sleep difficulties of the stressed and mentally diminished workforce and household. But Calm is only one of a growing number of partnerships held between UMG and digital health and wellness companies. Other partnerships include Thrive Global a "behaviour change product" that seeks to "drive productivity through wellbeing"; the neurotherapeutics company MedRhythms, which uses music as part of their efforts to improve walking, mobility, and other functional outcomes for those with neurological injury or illness; digital health company soundBrilliance, which aims to "empower people to better self-manage fundamental health—emotional balance, fitness, quality sleep & pain control"; and, of course, Vera, the app for dementia carers with which this book began.[64] In addition, as of 2024, UMG has launched their own "music-centric" audio-wellness venture, Sollos. The app provides music from UMG's catalogue, which is combined with "sound supplements"—that is, "special sound waves that have positive effects on your body, mind and soul": described as "audio vitamins," listening to these "science-powered sounds" promises to help listeners "focus, relax and sleep."[65] In its advertising literature, Sollos distinguishes itself from its competitors on the grounds that it is backed by cognitive science and psychoacoustics; that its audio is "artisanal," created by "human artists and producers" (albeit with the "assistance" of two proprietary algorithms); and that, while most "music designed to help you do things is boring" and consists of "AI-generated beats or bland, new agey spa music," Sollos offers music across a range of genres and for a variety of tastes.[66] Put simply, UMG's ownership of a huge archive of recorded music allows sleep to be soundtracked by a better-quality and a more "personalized" music selection.

The development of Sollos was announced alongside various digital health partnerships at a 2023 Music and Health summit, during which UMG's managing director Lucian Grange stated that he wanted the "powerful relationship between music and health" to be more than "a series of anecdotal observations": "I want it to be a key component of our strategy, so we can really put music to work in ways it has never been used before."

This strategy stemmed from UMG's "fundamental" belief that "music is a powerful force for good, and now we have the science and technology—with AI and therapeutics and more—to help accelerate these developments."[67] These partnerships suggest that putting music "to work" by enhancing a listener's physical and mental health provides valuable licensing opportunities: particularly as it seeks to develop accumulation strategies in relation to the AI investment boom.[68] And while it is currently unclear how Sollos will fare in relation to existing well-being apps and streaming services (including those to which UMG are attached), its announcement affirms that health and wellness is perceived as significant—and potentially lucrative—by the music company. Indeed, as of 2024, the global wellness industry has been reported as larger than the pharmaceutical industry and valued at around $6 trillion, its provision ranging from "luxury" retreats and "regenerative" medicine to comparatively inexpensive therapeutic commodities, such as (nominally) free digital services and self-help books.[69] As Suvi Salmenniemi highlights, its rise and expansion is partly symptomatic of the externalization of reproductive responsibility, and the accompanying drive to extract profit.[70] For those in need of care, "wellness" provides an alternative to austerity-stricken and potentially costly health services, allowing "autonomy" and "choice" through consumption. At the same time, by enabling investment in the self *by* the self (and the private household that sustains them), the wellness industry complements the neoliberal notion of human capital and its financialized modes of subjectification (see chapter 3).

It is no coincidence, then, that UMG are joining other large data companies such as Amazon in entering the digital health and wellness market (see chapter 2). As well as exemplifying a general trend for market consolidation and monopolization, UMG's licensing partnerships with Calm, Thrive, and soundBrilliance are symptomatic of a well-established convergence between the interests and activities of the recorded music industry and the digital technologies (or "tech") sector, of which much of the health and wellness market is now a subsidiary. Although much was made of the recorded music industry's crisis of profitability introduced by music's digitization and distribution on peer-to-peer networks, the introduction and adoption of internet circulation and digital technology as a means of creating and curating, sharing, grouping, selling, and licensing access to recorded music has been a significant catalyst within the music industries. Indeed, the establishment of the streaming sector—a key mediator between the music industry and digital tech sector—has been heralded as mitigating a two-decade revenue decline for the recorded music industry, enabling the "peak" of the compact disc market in the late 1990s to be exceeded in the 2020s.[71] Other effects of the convergence are traceable in a shift whereby recorded music has become content as opposed to a product, with recording artists acting as content providers as opposed to creative producers.[72] The abundance of audio files available on digital platforms, furthermore, place renewed emphasis on the "curator." These observations are recognizable from Calm's app. While its audio library is relatively compact compared to streaming services such as Spotify—something that some commentators have suggested may be attractive to artists insofar as songs in apps become highly targeted and discoverable[73]—users nonetheless face an abundance of audio. They can navigate this content using various tools, including the "discover" tabs, where users can access audio by type (e.g., soundscapes, music), function (e.g., sleep, meditation, movement), listener (e.g., kids), or length of time, and recommendation algorithms that suggest new content based on previous listening activity. In addition, users can access "curated by Calm" playlists, consisting of tracks by multiple artists, or "infinite tracks," which are described as being "designed and created by real musicians but built to play infinitely and be ever changing."

Although the changes in music production and dissemination resulting from the convergence of music and digital tech sectors should not be downplayed, it is also important to note that there are structural precedents and historical continuities that should temper perceptions of the rise of the digital technologies and their associated market sectors as generating radical disruptions and transformations in music cultures and economies. As Dave Hesmondhalgh and Leslie Meier make clear, the activities, structures, and outputs of the music industries have long been impacted by their intersectoral relations with other industries. It is notable, for example, that the record industry emerged from the phonograph industry, with established institutions of the music industry showing little interest in investing in the early record business.[74] Likewise, prior to the early twenty-first century, consumer electronics companies were a key force in shaping how music was mediated and experienced, with recorded music often functioning—from the perspective of their parent companies at least—as a means of attracting consumers to new technologies.[75] This dynamic has persisted; however, it is now telecoms and IT companies who are shaping recorded music's mediation and consumption. The interests and investments of these companies are traceable in the transformations of music into downloadable MP3 format (for which Apple's iPod and iTunes store was a key player and beneficiary) and then streamed content, emerging with the establishment of cloud computing.[76] While the pace of change relating to music's mediation and consumption in the twenty-first century is striking, and the wide-ranging musical, social, economic, and ecological implications of recorded music changing from something possessed to something (conditionally) accessed should not be downplayed, there is nothing particularly new in the convergence between recording companies and nonmusic sectors.

Although apps such as Calm make clear the ongoing mutual interests of the recording music industry and digital tech sector, convergence should not be mistaken for coherence: their relationship is rather one of antagonism and dependency. Partnerships often occur alongside conflicts: the recording industry's disputes with YouTube and other streaming services around 2016 regarding remuneration to rights holders is instructive in this regard. As Keith Negus argues, these conflicts were symptomatic of a schism relating to differences in business models: The recording industry was largely attached to a model consisting of investment inartistic production and remuneration through copyrights and unit sales. Digital conglomerates (e.g., Google) and their associated platforms (e.g. YouTube), meanwhile, generate income from the way "content" attracts advertising (which has predecessors in the use of advertising to finance commercial radio and television). For Negus, in the latter model, in which music appears as "content" for attracting subscriptions and traffic, the artistic qualities and the effort that goes into composition and production are largely irrelevant. This, and the limited financial value of music compared to apps and audiovisual content, means that large data conglomerates tend not to invest in the production of music. However, this limited expenditure on music should not be mistaken for its insignificance. As was made clear in chapter 2, digital conglomerates exploit music as an intimate means of data accumulation. This data has various uses, from driving recommendation algorithms, predicting consumer behavior, and analyzing user engagement to generating revenue through its sale to advertisers.[77]

The case of Calm both affirms and offers a revealing counterpoint to Negus's analysis. While the specifics of the process are opaque, it can be assumed that Calm employs music as part of its collection and collation of user data, which it uses to "personalize your online experience and the adverts you see on other platforms."[78] Calm also collates its own data with data from other sources, including marketing and advertising partners, consumer research platforms, business contact databases, and publicly available sources of information, and shares "aggregated information" with third parties.[79] The company has credited its use of Amplitude—a data analytics software company—as enabling the company to make "smart bets and risks" in developing new initiatives and to track patterns in user behavior.[80] Music, then, is one type of "content" that feeds into the production and collation of data and, by extension, the generation of both advertising revenue and potentially lucrative user insights. If, as Robert Prey suggests, "all listening-time is data-generating time," then this suggests a complementary relationship between the economic models of digital tech companies and the provision of audio for activities such as sleep, which typically require the user to "listen" (or, rather, "play audio") habitually, and/or for an extended period.[81]

At the same time, Calm's functionality as a sleep, meditation, and wellness tool means that the company *does* have some interest in music's artistic qualities. As noted previously (and despite the audible aesthetic homogeneity of the app's audio library), Calm has positioned itself as an avenue of creative freedom for artists, allowing them to experiment with creating longer-form pieces. The rhetoric of the company's executives certainly offers an interesting subversion of the usual relationship between musical functionality and aesthetic autonomy. Calm's head of music, Courtney Phillips (who herself previously worked at UMG) has suggested that the requirements of the app work against the commercial imperative to cut down music to make it more marketable: in an interview with Rolling Stone she states that she wants musicians' "weird side projects and meandering pieces."[82] In striking contrast to the formulaic associations of Muzak, for example, or indeed the "new agey spa music" derided by UMG's

Sollos, Calm posits the production of music for sleeping as an opportunity for transcending aesthetic constraints: compared to the "formulaic" three-minute pop song, the platform's musical workers are relatively "free" from market-oriented restrictions, allowing them to find fulfilment (and potentially even remuneration) in pursuing their experimental (i.e., weird, meandering) "projects." Furthermore, Calm has directly invested in music's production, insofar as it has commissioned new music by well-known artists, with the platform holding exclusive distribution rights: there is, for example US violinist Lindsey Stirling's "Lunar Lullaby" and country singer Keith Urban's "I am Home." However, the precise terms of these commissions are unknown, and it is unclear the extent to which Calm takes on the financial risks associated with creating new music: a risk that, historically, has been taken on by record labels and companies or by artists themselves. Nonetheless, Calm, unlike the digital companies examined by Negus, appears to have some involvement and investment in music's production: music remains a means of attracting subscribers and advertisers, as well as a means of data accumulation but it is also central to what the app promises to do. If sleep requires the creation of audio content with a particular aesthetic—not all audio is "sleep audio"—then there remains a degree of investment as to what is produced for and included in the app's sound library.

In a presentation for the tech speaker series Google Talks, Calm's cofounder Michael Acton Smith, responding to a question about his own beneficial sleep habits, states that he tries not to use his phone in bed. Reaffirming that the crisis of sleep is a crisis of (digital) distraction, Acton Smith suggests that he resists turning his phone on until he leaves the house, and this "30, 40 minutes [of] breathing space" gives "a powerful foundation for the rest of the day."[83] The implied contrast between Calm's mission and the distractive dynamics of digital space is made explicit by Calm's marketing executive, Sue Cho, who observes (in a talk entitled "Breathe Deeply and Disrupt") how "in today's digitally fuelled world" companies and channels fight for attention, encouraging users to keep scrolling, consuming and doing more. The point is accompanied by video clips of phone screens showing multiple tabs and notifications. Calm's objective, by comparison, is "stop scrolling and do nothing."[84]

These claims by Acton Smith and Cho, which draw on popular rhetoric about the anxiety-inducing effects of social media and smartphones, could simply be dismissed as predictable hypocrisy: Calm is yet another product of today's "digitally fuelled world" and is primarily accessed through smartphones. But they also speak to a series of wider contradictions that underpin Calm's sound-based sleep aid, whereby phones are positioned as both an inhibitor and producer of rest, and where Calm, whose economic model is reliant on data accumulation, needs to get the user to do something as they "do nothing." These apparent contradictions are not so much resolved as obscured due to Calm's status as an *auditory* digital service. When using the app, the phone becomes something to listen to rather than look at: although initially accessed via the screen, the speaker takes over as the means of engagement. As I discussed in chapters 2 and 3, the auditory has been frequently construed as intimate, intuitive, and (paradoxically) unmediated: various cultural ideas and ideals present aurality as a "natural" and "primary" mode of sensory engagement. These constructions help to figure the smart speaker's surveillance as service (see chapter 2). Likewise, in the case of Calm, they both cohere with its "soft" auditory aesthetic—voices are close miked, music is unobtrusive—and complement sound and music's usage as a tool for generating rest *and* data. Consequently, "listening" to music presents as "doing nothing."

## RENEW: THE DE/COMMODIFICATION OF MUSIC

In beginning with the lullaby and ending with Calm's musicalized wellness app, this chapter might be understood to present a narrative in which social reproduction and music, and the relationship between them, have been subject to commodification. As a form of reproductive musical labor, the lullaby appears as one of the "free gifts" stemming from naturalization of feminized care, requiring neither remuneration nor resource. It can also be understood as a type of common good. The lullaby is not owned but shared, passed around by kinships and communities, existing predominantly as a folk tradition rather than as privatized property. There are exceptions, such as the compositions stemming from Carnegie Hall's Lullaby Project. Nevertheless, the lullaby remains a shared musical resource.

In contrast, sound-based sleep aids, which feature music alongside other types of audio, are commodities that tend to be bought and sold as either goods (technological artifacts that can be used again and again) or digital services (as is the case with subscription-based products whereby users purchase access for a defined period of time).[85] These devices capitalize on the relationship the lullaby draws between music and sleep, turning it into an opportunity for monetization and profit generation. Their rise seemingly mirrors the subsumption of other socially reproductive activities. Just as ready meals can substitute cooking, and buying clothes can substitute making and mending them, a speaker device with an inbuilt audio library or an app installed on a smartphone can stand in for the lullaby. However, the commodified sleep aid also expands the lullaby's listener demographic, insofar as working adults are presented as one of, if not the primary beneficiaries of these technologies.

None of this is to suggest that sound-based sleep aids have simply replaced the lullaby: contra the dystopian image of "The Robot's Lullaby" (see chapter 1), people still sing lullabies to encourage infants and children to fall asleep, just as they still cook and make and mend clothing.[86] The lullaby persists, but the rise of the sound-based sleep aid is itself telling of various social, economic, and technological transformations. As a result, both music and sleep are "put to work," to echo Universal's Lucian Grange.

There are, however, important factors that complicate this narrative of commodification, with which music for sleeping becomes market mediated. These complications arise because apps (a key subsection of the sleep aids available on the contemporary market), like music itself, do not always have a straightforward relationship with the commodity form. The types of exchange and exploitation that ground the political economy of apps, streaming services, and the internet more widely have been subject to increased scrutiny, with some commentators suggesting that a model of rent extraction and asset acquisition more accurately describes the transactions between users, app-based digital services, and advertisers, occuring alongside—other types of exchange grounded in commodity production (e.g., the production of smartphones, headphones, speakers; software design and maintenance).[87] Indeed, app-based digital services can encompass various types of transaction insofar as they enter into exchanges with multiple parties: the relationship between user and app is one part of the story. As Calm's partnership with Universal attests, there are other exchanges that occur between various economic actors, with music licenses being bought and access to accumulated data and advertising space being sold. From the perspective of the user, the payment of a subscription fee does not amount to the purchase of music: it does not result in ownership of a musical commodity. Rather, it is a means of securing access to the "asset" of the app's (or another digital service's) music library. It is the digital service that purchases music (or rather, music licenses) to produce the library-as-asset as well as one-off file copies (i.e., streams). Consequently, as Rasmus Fleischer, among others, has observed, music is both commodified (in the sense that music licenses are purchased by the streaming service) and *de*commodified (in the sense that, for the listener-consumer, the individual song no longer has a price).[88]

Another potential challenge to the narrative of commodification is the plethora of sound-based sleep apps that are available for free: if a commodity is defined by use-value and exchange-value, then, from the perspective of the user, free apps have no exchange-value. Higher-end apps such as Calm offer free trials and limited "freemium" content, however, users are repeatedly pushed by the platform to sign up for a paid subscription. However, if companies such as Calm are at the top of the pyramid that is the sleep aid market, towards the bottom there is an abundance of free apps, many of which appear very similar in terms of design, function, and features. In some instances, the app's free status is reflective of its quality, reputability, and functionality. A significant proportion of these apps have very few users and are associated with developers with no or little public profile. Many also have performance issues: poorly placed advertisements can make it difficult to interact with the interface, audio loops may be glitchy, or playback may stop suddenly without explanation. These software products are by no means exceptional anomalies: they are, like spam, viruses, malware, broken weblinks, and, more recently, "AI slop," common features of networked culture. App stores have tried to take action to limit the availability of "junk apps" or "crapware" so that searches do not produce hundreds of results that are difficult to distinguish from one another.[89] Yet problems still remain: a January 2024 search of the Google Play app store using the phrase "sleep music," for example, returned hundreds of results, more than thirty of which included the phrase "sleep sounds" in their title, featuring a similar thumbnail image of a crescent moon and stars with a blue and purple color scheme.

In instances when use of an app's service is free, the "cost" of free access is intended to be offset by other income streams (e.g., advertising revenue, surplus generated by paid subscriptions). To this end, it is exchanged for something other than money—namely, the user's capacity to pay attention to advertisements and access to user information. In other words, the app's audience are a resource, providing both data and the possibility of attention which are, in turn, transformed into a commodity or asset. However, it is important to note that while data accumulation and user attention are key to the economic model of digital services, in practice, they are often unsuccessful. The "space for ads" box in the middle of the interface of Rainforest Media's Sleep Sound app (figure 9) highlights that in a saturated market, it can be difficult to attract advertising revenue, particularly for apps with small audiences and relatively unknown developers. Yet even for bigger companies, advertising revenue may become increasingly difficult to sustain, with the growth of ad blocker services as well as bot-generated user views and spam.[90]

Despite these complexities, the rise of the sound-based sleep aid coheres with other technologies discussed in this book, in that it promises a technological fix to a "problem" of life making, a sonic surrogate for the musical labor historically performed by feminized caregivers, which brings with it the prospect of commercial gains. As an outcome of the establishment of the sleep industry in the late nineteenth century that capitalizes on an apparent crisis of sleep, the sound-based sleep aid first becomes a means of expanding commodity consumption, taking the form of a household electrical appliance. Moving into the twenty-first century, the sleep aid then also became an app and a participant in both the digital economy and the music industry—a transformation fueled by a financialized drive for musical and technological interventions into health and wellbeing. Where many sleep apps appear to make use of royalty-free library music, some app companies have taken on more active roles in commissioning and curating new music, working in partnership with the industry's major players. For record companies, meanwhile, a renewed emphasis on music's functionality in relation to sleep, but also relaxation and mental well-being, presents an opportunity for expanding revenue

*Figure 9.* Sleep Sound app home screen.

streams at a time when streaming has become a—if not the—dominant means of consuming music. UMG's development of their own in-house music well-being app is a testament to the perceived value of such uses of music. However, if the rise of the sleep aid means that the domestic lullaby has been partially exchanged for professionalized musical labor, the extent to which musicians themselves benefit from this exchange remains unclear. While apps such as Calm present themselves as providing opportunities for creative freedom, there is very little information on how these opportunities are remunerated.

## CONCLUSION: FINDING THE DREAMSPACE

Where does sleep's capture by capital leave aspirations for a liberatory politics of sleep? The emergence of the sound-based sleep aid—first as product, then as digital service—certainly tempers the claims of those such as Crary, who posit sleep as an autonomous, "unassimilable" practice: a break from the "constant continuity" of a 24/7 work culture. The sleep of sleep apps affords no such break: it is both put to work and put to the service of work, insofar as improving sleep is associated with improving productivity and performance, ensuring that one can "keep going" under mundane and stressful conditions. Sleep, then, is structured by the requirements and temporalities of production. Yet Crary should not be dismissed too hastily. Sleep can, after all, be mediated by other ambitions and demands. As a cultivated practice, it can be put to use for other ends.

Sleep has been harnessed as a mode of dissent and direct action, pursuing what Neil Smith speculatively describes as sleep's "quintessential social inventiveness" and "political creativity."[91] Such was the case when, in spring 1971, Vietnam Veterans Against the War, collectively fell asleep on Washington's National Mall. As Franny Nudelman outlines, sleep became an integral part of the group's protest after a court's injunction: the veterans were permitted to stay overnight in the Mall and were allowed to sing, sit, or even lie down. Sleep, however, was legally prohibited. In defiance of the order and the sleepless miseries of war, hundreds slept in protest.[92] More recently, the collective feminist action "Meet to Sleep," first conceptualized by Jasmeen Patheja in 2008, seeks to reclaim "the right to be defenceless" by bringing together participants from across India to sleep in public spaces.[93] Meanwhile, artist and theologian Tricia Hersey, inspired by Womanism, Afro-futurism, somatics, and Black liberation theology, has interrogated the connections between rest, reparations, and resistance through performance art, community organizing, and installations. Under the rubric of "the Nap Ministry," Hersey's work often centers the trope of napping, which she posits as both a spiritual practice and a crucial means of resistance for those marginalized by racism, poverty, violence, and discrimination. For Hersey, napping, alongside other modes of rest such as daydreaming, meditation, and the enjoyment of leisure time, can be mobilized to help recapture the "dream space" stolen by slavery, plantation labor, and the capitalist commodification of Black bodies.[94] Sleep is resistive in the sense that it disrupts the imperative to "do more" but also in the sense that it provides opportunities to imagine alternative modes of being and energizes social struggle.

The sonic has been a recurring component of Hersey's attempts to cultivate a radical praxis of sleep. The Nap Ministry's "Nap sessions" are events that allow communal resting and typically feature music, sometimes performed by live musicians.[95] She has set up a telephone hotline (1-833-LUV-NAPS) featuring recordings of Hersey reading "rest messages," as well as poems and excerpts from theological texts. In 2021, she released Rest Life: a rest meditation in the style of a slow jam. The track opens with the words "welcome to the dream space," a line that becomes a refrain. A groove constructed of electric guitar, synths, trap hi-hats, and full bass underlines spoken affirmations: "Our collective rest will save us"; "Imagine a world where capitalism doesn't exist, poverty is no longer created."

While superficially similar in using music to cultivate rest and renewal, the political Blackness of Hersey's rest meditation puts it at a significant aesthetic and material distance from the ambient sleep songs of Calm and similar sleep apps. Instead of stasis, there is rhythm. Self-care is inextricable from collective care.[96] Sleep is not for the revitalization of productivity but a site of refusal and speculation. Transformation rather than perseverance is the aim. This is not to suggest that Rest Life, or other aspects of Hersey's project simply exist outside of capital: Rest Life was released via streaming platforms, while Rest is Resistance easily circulates on social media as a compelling slogan, attracting potential purchasers to a best-selling book of the same title. Indeed, the success of Hersey's work could be considered symptomatic of its partial compatibility with a digital "influencer" and "wellness" culture. Yet it is from the inside that an escape is planned, one that harnesses the instability at the core of reproduction as both the means of accumulation and its embodied limit. Here, music helps to draw rest into conflict with capitalist temporalities rooted in the plantation and to bring into audition the necessity and possibility of fulfilment: "Welcome to the dream space."

# Epilogue

## REFUSING REPRODUCTION

In 2022, TNFW Nique's audacious "BDF (Baby Daddy Free)" began to reverberate across the internet, generating outrage and delight among listeners. In this remix of GloRilla and HitKidd's viral crunk anthem "FNF (Let's Go)," a two-minute song about the freedoms and pleasures of being a newly single woman, the Oakland rapper maintains the ominous keyboard loop, strong beat, and infectious defiance of the original, while adapting the lyrics. Narrated from the perspective of a woman seeking an abortion, "BDF" speaks to a desire to remain child-free and unattached. Riffing on anti-abortion hyperbole, the song's protagonist describes herself as having "murder" on the mind: this becomes a call that finds its response in the signatory "let's go!" vocal hook of "FNF," hailing listeners to sing (or shout) along. The video for "BDF" similarly references and modifies that of the original. Both videos depict women partying together in and around a car in a parking lot. Yet where "FNF" features GloRilla partying with her friends in the daytime, "BDF" is shot at night, with TNFW Nique and a dancer twerking and throwing money outside a Planned Parenthood building.

Listening again a few years after its release, "BDF" takes on new resonances. Around the bay from TNFW Nique's Oakland, pronatalist politics have gained traction among Silicon Valley venture capitalists seeking to secure a future amenable to their own interests. Driven by the specter of population decline and its destabilizing social and economic effects, its Silicon Valley advocates have promoted pronatalism as a means of preventing the end of "civilization" and uplifting the human species, the nation, and the world. In so doing, they have contributed to a broader right-wing coalition that, despite key ideological and moral differences, finds unity in gender essentialism and a desire to control reproduction.[1] Just as "family values" enabled alliances to be fomented between neoliberals and neoconservatives, the desire for "more babies" brings together the tech bros with trad Christian conservatives. In the US, this potent yet uneasy alliance has coincided with, informed, and justified the rescindment of reproductive and gender-affirming health care, executive orders concerned with reducing the costs of assisted reproduction treatments, and the proposed direction of resources toward communities with higher marriage and birthrates.[2] It is also helps fuel a revanchist cisheteropatriarchal politics, traceable in the demonization of queer and trans communities, as well as the renewed proliferation of scientific racism and eugenics.[3] Indeed, the dovetailing of pronatalism with an anti-immigration and ethnonationalist agenda is telling: when the problem is described as birth rates, what is often meant is white birth rates.[4]

To be clear, the specter of population decline is not fabricated whole cloth by pronatalists. Fertility rates *have* been falling over the course of decades, and many parts of the world are now "post-reproduction" societies.[5] There are various social forces behind this trend, including greater availability of contraception and the greater number of women in work, but the rise in precarious employment and wage stagnation play a role too. Indeed, Paul Cammack suggests that the growing scarcity and insecurity of work that is characteristic advanced capitalist development means that there is a structural "tendency for the rate in fertility to fall" in capitalism.[6] Significant, too, is the lack of social provision for childcare, health care, housing, and education in many countries, the real and prospective effects of climate crisis and geopolitical instability. While individual decision-making also informs procreative choices, wider social, economic, and environmental conditions feed into a trend whereby reproduction is collectively refused.

In this context, "BDF," too, appears as refusal, albeit one that is both complicated and intensified by anti-Blackness. The racial and reproductive economies of chattel slavery, which continue to shape social life and death within and beyond the US, have characterized Black women as abundantly fertile, hypersexual, and maternally irresponsible; subjected them to forced sterilization, obstetric violence, and state surveillance; omitted them from the institution of motherhood and the (limited) protections it may afford; and disproportionately exposed them to harm in pregnancy and childbirth.[7] The events driving the Black Lives Matter movement has reaffirmed Black motherhood's proximity to both Black life and death, as both protectors and grievers. Stemming from this, as Jennifer C. Nash argues with reference to the rhetoric of the US left, the "long-standing conditions of the ordinary" are newly framed so that Black mothers become emblematic of crisis.[8] Affectively constrained by the need for respectability, the symbolic currency of the Black mother affords moral superiority to those who evoke it, all the while becoming another weight to carry by actually existing Black women, reinforcing the notion that Black motherhood is a problem in need of intervention.

If, as Nash contends, the post-BLM left has selectively foregrounded the suffering associated with Black pregnancy, childbirth, and mothering, so too have elements of the anti-abortionist right, who have invoked the racist histories of reproductive healthcare for their own ends. TNFW Nique's performance of "BDF" in front of a Planned Parenthood, for example, is striking given the organization's recurrent casting as limiting Black births. In 2023, for example, the Ohio Republican Senator Michele Reynolds decried abortion as a form of "extermination" for the Black community, describing Margaret Sanger, the founder of Planned Parenthood as "a eugenicist who systematically targeted 'inferior races' for decades."[9] The 2016 Republican presidential candidate Ben Carson similarly supported a 2015 campaign to defund Planned Parenthood, claiming (without substantiation) that Planned Parenthood had most of their clinics in Black neighbourhoods and Sanger, as its founder, had not been "particularly enamoured with black people."[10] Sanger was certainly proximate to the racist and ableist ideological project that was the US eugenicist movement. At the same time, as Sarah Whitmarsh notes, "There are no primary sources that cite Sanger's [direct] support for race-based eugenics policies": Sanger is "a woman whose legacy is fodder for both accurate, justified criticism as well as false or distorted rhetoric."[11]

When Black conservatives such as Reynolds and Carson present their opposition to abortion as a stand against population control, they inevitably leave unanswered a question hinted at by the refusal of "BDF": what might be necessary for reproductive *justice* to be secured *for all*?[12] Stemming from the intellectual labor and activism of women of color feminists, reproductive justice counters a constraining focus on abortion access and personal choice to address the wider social and economic conditions that shape procreative decision-making, often working in coalition with movements for self-determination, antipoverty, and antiracism toward a liberatory horizon of sexual, bodily, and reproductive autonomy.[13] From this perspective, Carson acted against reproductive justice when he claimed that poverty "is a state of mind" or when he advocated for removing public housing protections and funding cuts when he headed Donald Trump's Department for Housing and Urban Development.[14] Reproductive justice, then, shares the approach taken over the course of this book, since it necessitates thinking and addressing pregnancy, gestation, and childrearing in tandem with broader social and economic relations that enable or constrain the means of life making and in contradistinction to notions of natural proclivity and personal choice. Such a framework is gestured toward when "BDF" connects abortion seeking with milk shortages and diaper prices, and with a desire to remain outside the gendered constraints and demands for care that accompany the heterosexual couple form: having a baby is presented as neither economically viable nor socially desirable, and, under such conditions, reproductive justice remains unrealized. The point here is not to give substance to a pronatalist standpoint by asking what it would take to convince TNFW Nique to make a different choice. Nor is it to affirm, in its negation, Black motherhood as a spectacle of suffering. Rather, it is to recognize the complex historical, social, and economic background conditions against which the protagonist of "BDF" boldly announces her refusal to partake in gestational labor, instead choosing the pursuit of pleasure without apology or constraint.

Much of this book has focused on the affirmation of reproductive respectability and responsibility via sound and music. Instrumentalized as a means of automating and assisting in the work of care, life making, and maintenance, recorded music and its technologies of mediation speak to reproduction's social necessity and its coconstitution with, to borrow the phrasing of the Endnotes collective, "gender, race, class and other misfortunes."[15] These "promissory" sonic surrogates are reliant upon both the constitution of the auditory as an intimate, intuitive, and affective medium and recorded music's production as a cheap and abundant resource, allowing the mother to choose the right record for her children, the smart speaker user to "just ask," or transport companies to produce hours-long playlists of "repellent" classical music. They are also emblematic of efforts to use recorded music to drive down the costs associated with other types of provision. It is cheaper, after all, to purchase a smart speaker and Amazon music subscription than it is to pay for home help. It is cheaper to entertain the household with a phonograph than it is to employ a nanny or a piano teacher. It is cheaper to offer a music-based well-being app than it is to pay for therapy. It is cheaper to broadcast unwanted classical music than it is to hire security.

As chapter 1 attests, the promise of recorded music's utility with reference to the making and the maintenance of the family can be traced back to the turn of the twentieth century, when the mechanization of music via early sound recording and reproduction also enabled music to *become* mechanization, supplementing, standing in, and transforming the gendered labor processes of the home. The notion that recorded music can help us take care of ourselves and others for whom we are obliged to care is as old as the recorded music industry. At the same time, the proliferation of reproductive sound technologies from the 1990s onwards is expressive of some of the social and economic reconfigurations associated with post-Fordism, neoliberalism, and financialization and with the commodification and externalization of reproduction. Rent-extractive sound-based sleep aids, below-cost smart speakers, the "sound investments" of prenatal soundsystems, and the surveillant listening of revanchist noise-reporting apps are all conditioned, in one way or another, by the contraction of the formal economy, the erosion of the social wage, the collective degradation of living standards, and the unprecedented centrality of debt to capital accumulation. Ultimately, reproductive sound technologies are symbols of capitalism's privatization of need: the separation of people from their means of reproduction, necessitating that one "work to live," whether that work is available or not.

If music, like care, is often uncritically valorized for its subjective, social, and ethical benefits, then to attend to its status as a reproductive sound technology is to grapple with music's expansive role in maintaining capitalism's fundamentally unequal, oppressive, and exploitative means of life. Even as they both remain partially external to market logics, music and care are ultimately constrained by the totalizing drive for accumulation, as well as the crisis tendency inherent to this drive. It would be a mistake, however, to reduce music *or* care to the affirmation of capitalist life. "BDF," to this end, is emblematic of the pressure music exerts upon the reproductive norms of capitalist society, by refuting, contradicting, or imagining a life outside of them. It belongs to a substantive, knotty tapestry of popular music that gives voice to frustrations about the nuclear family and the gendered labor it demands, as well as making felt the joys that lie at its margins. Musical tropes around the satisfaction found in exclusive, heteronormative love are internally countered by the rich body of song that expresses its disappointments and instead finds satisfaction external to it. If music idealizes the heteronormative couple form, then it also provincializes it through its manifold, transgeneric expressions of sexual desire that take place outside its frame of reference.

Connected to but distinct from this is popular music's recurrent invocation of the pleasurable solidarities that can be found and felt outside the nuclear family: whether that takes the form of "getting wild" to invoke GloRilla's "FNF," the "peas in a pod" of Nadia Rose's "Skwod" (which became the unofficial anthem of UK activist group Sisters Uncut during an antifascist mobilization in Croydon[16]), or the unconditional offer of a place to stay made in Aaliyah and Missy Elliott's "Best Friends." As these examples elucidate, Black popular music has placed particular emphasis on non- or extrafamilial communities of care, occurring as a trope across the progeny of R&B and hip-hop. This is no accident: if the logic of the couple form is concurrent to the wage-form, then it is unsurprising that those who have historically existed at a remove from the latter's (limited) benefits also find themselves detached from the (limited) benefits of the former.[17] Granted, the "friends," "sisters," and "girls" tend to show up in song once the couple form has proven itself to be a letdown, appearing as a safety net that comes into action in times of romantic crisis, the shock absorber of the betrayal and hurt that accompanies heteronormative, familial love. Yet in its sounds, lyrics, and videos, popular music also recurrently hints at the joyous fulfilment that might be found in these relationships, enabling care to be experienced as abundant rather than scarce. While the common designation of popular music as escapism—one which speaks to its alignment with capitalist assignments of leisure time—makes it easy to dismiss the desires, imaginaries, and worlds implicated in it, it matters that popular music is a cultural site in which the failings of capitalist coupledom *and* the possibility of a sustainable and fulfilling life outside of it can find sensuous expression.

There are other ways that music exerts pressure upon capitalist modes of care. This includes the alternative—and sometimes oppositional—forms of community, sociality, and modes of production that are fostered through music. The alternative economies and experiments of being associated with electronic dance music, punk, and various grassroots "do it together" scenes have sometimes lent themselves to uncritical romanticization and nostalgic revisitations that downplay the tensions and contradictions inherent to their organization.[18] Yet while they tend to remain small scale and are often rendered precarious by the dual forces of capital and the state, their prefigurative experiments in resource sharing and cooperation can serve to denaturalize the market logics of other musical and social worlds. Furthermore, the production and performance of music is entangled with important communal efforts to foster comfort, safety, and mutual aid for and among those who, due to racism, cisheteropatriarchy, and ableism, are estranged from capitalist modes of life making reliant upon the wage, family, and private property. It is perhaps unsurprising, for example, that wider cultural interest in the caring "homes," self-sustaining social networks, and "dream-space" (see chapter 5) provided by ballroom culture has flourished at a time when social precarity is expanding.[19]

There is also music's troubling of reproductive time and its capacity to disrupt the temporal logics of work, rest, and replenishment. "The night time is the right time" states Paul Gilroy, riffing on the popular R&B song title. Gesturing to a specifically Black form of critique of capitalist time and space from the position economic liminality, Gilroy identifies how in bars, clubs, and dancehalls, "the period for recovery and reproduction is assertively and provocatively occupied instead by the pursuit of leisure and pleasure.[20] Here, we might recall the nocturnal staging of TNFW

Nique's BDF, which also, in its invocation of stripper culture, hints at the gendered dynamics that accompany the pursuit of these pleasures. However, Gilroy was writing in 1987: as acknowledged in chapter 5, the temporal boundaries separating work and rest have become increasingly fluid and unstable. This has been accompanied by the formalization of music's nighttime economy, as well as the expansion of music consumption to accompany all of life's activities, including sleep. While this might diminish music's disruptive charge, it does not entirely scupper it: from *pancadão* (see chapter 4) to the disorientating scenes of daytime revelry depicted in GlorRilla's "FNF," the free time of musical enjoyment may still create friction with the imposed times of work and rest.[21]

Pertinent, too, is music's involvement with more pronounced forms of activism confronting the assignment of disposability to vulnerable populations. We might consider here music's transnational imbrication with grassroots activism related to the HIV/AIDs epidemic: from the care programs run out of nightclubs to the use of song and performance to offer counter-narratives, share health information, and advocate for access to antiretroviral therapy.[22] Or consider music's involvement in antigentrification struggles and housing activism, both of which were at play in the Don't Mute DC protests, sparked by a noise complaint lodged against a shop that had long played Go-Go through a speaker outside its premises. Likewise, a 2024 action at Tribulate 7 in Lavapies, Madrid, saw musicians perform in living rooms, balconies, doorways, and kitchens to mobilize resistance against the building's purchase by a real estate investment trust and the prospective mass eviction of its residents.[23]

It would be misguided to present music as offering a clear solution to the problem of capitalism's social forms. Contra the proliferation of reproductive sound technologies, music cannot provide us with a cheap and easy fix for the crisis of reproduction. Nor can it provide us with a cheap and easy fix when it comes to fighting for the reorganization of life making to allow our collective flourishing, something which ultimately necessitates nothing less than the deprivatization of need: music cannot automate struggle any better than it can automate care. At the same time, the antagonistic resonances that stem from music and the myriad modes of production, circulation, sociality, and expression contained within it, should not be too quickly discounted. Indeed, they speak to a significant formal affinity: music, like reproductive labor, often moves to a different beat to that of the wage.[24] It is from this position that music both affirms and disavows reproductive respectability, confirming its limits but also sounding out alternative horizons of pleasurable fulfillment without cost.

# Acknowledgments

Writing this book has been a long and slow process, during which there have been many life changes. I am indebted to many colleagues, comrades, friends, and loved ones who have helped and held me along the way.

It is undeniable that underpinning this book is a profound dissatisfaction with the present and the prospect of a "competitive future." Much of my adult life has occurred under a seemingly unending period of austerity, and many of the issues that frame this project—the growing instability of waged work, the increased pressures in producing a "liveable life," the slow rescindment of social provisions—are by no means mere abstractions. My thinking about music's deployment as a "caring" technology has been conditioned by the difficulties so many of us face in securing—and providing—adequate physical, psychic, emotional, social, and reproductive care. I want to acknowledge the myriad conversations and circumstances that have, in one way or another, brought clarity in affirming that care is of vital importance and, presently, woefully inadequate. I nevertheless remain inspired by (and have benefitted from) the networks of care that exist despite and because of these inadequacies. Thank you to all of those who struggle for our collective fulfillment.

This book began to take form during a period of research leave in 2019 at The University of Edinburgh's Institute for Advanced Studies. For this, I am grateful to my sponsor, the late Rebecca Collins: a brilliant listener, artist, and thinker whom I miss a great deal.

I was fortunate to have such supportive colleagues while working in (what was) the University of Lincoln's School of Film and Media: thank you in particular to Emily Wilczek, Sandra Kazlauskaite, David McSherry, Gábor Gergely, Rob Coley, and to the many inspirational students I met during my time there. While writing this book, I joined the Music Department at The Open University, and I am grateful to my more recent colleagues for making me feel so welcome in the discipline and at the institution. Special thanks goes to Helen Coffey, Byron Dueck, and Martin Clarke who have covered some of my responsibilities while working on the final stages of this book.

Thank you to Jean Ma, Jim Buhler, Raina Polivka, Sam Warren, and the rest of the UC Press team for their enthusiasm and support for this project; and thank you to Jon Dertien and the team at BookComp, Inc. for their editorial work. Thank you to the editorial collective at Feminist Review, who published an earlier, shorter version of chapter 3. Thank you to Dayna McLeod for your work, and for allowing me to reproduce images of Intimate Karaoke, Live at Uterine Concert Hall. My thinking has benefitted from conversations with Lílian Campesato, Ruth Charnock, Antoine Chessex, Marcelo Conter, Adam Denton, Marren Haffke, Miranda Iossifidis, Frances Morgan, Pedro Oliveira, Greg Seigworth, and Thibault Walter, among others. Various sharp readers have offered feedback, encouragement, and constructive criticism: thank you to Clara Latham, Paul Rekret, Eric Drott, Steff Doktor, Thao Phan, Marianna Ritchey, and Tom Sutherland for all your insights.

Thank you to Annie Goh and to all those who I have found myself thinking, doing, and making with as part of Sonic Cyberfeminisms, and to the amazing Olivia Hellewell, with whom I have shared the difficult final stages of the writing journey. Thank you to Nottingham friends for making this city home, and to Helen and Ale for being an important presence in my life despite being further away.

I have benefitted to no end from having a loving family that makes my life lighter: thank you to Christine, Adrian, Claire, Craig, and Harriet for everything. I recognize in some of what appears here the positive influence of my Nanna, Sheila Rose Thompson, who passed away in late 2025. I owe a great deal to my partner David, who has given so much over the years. Thank you for your love, encouragement, kindness, and cooking.

Finally, to Rowan, who came into this world as I was writing this book and turned my life upside down in the best possible way. Thank you for teaching me the joyous creativity of care.

# Notes

## INTRODUCTION

1. Booth, "12 Self-Care Playlist Ideas to Listen to When You Just Need Time for Yourself."

2. Paine, "UMG Partners with Music Wellness App Vera for People Living with Dementia."

3. This description is given in Paine, "UMG Partners with Music Wellness App Vera for People Living with Dementia."

4. The economic barriers to accessing dementia care and the impact upon family caregivers are discussed in Hutchings et al., *Dementia: The True Cost; Fixing the Care Crisis*. The gendered expectations regarding dementia care are discussed in Alzheimer's Disease International, *World Alzheimer Report 2019: Attitudes to Dementia*.

5. A clear articulation of this narrative can be found in Fraser, "Contradictions of Capital and Care."

6. Endnotes, "The Logic of Gender"; Fraser, "Contradictions of Capital and Care."

7. The policing of reproduction is discussed further in chapter 4. See also Mies, *Patriarchy and Accumulation on a World Scale*; Wang, *Carceral Capitalism*.

8. These trends are discussed throughout the book. For more on streaming and its impact on the music economy see Drott, *Streaming Music, Streaming Capital*; Fleischer, "If the Song Has No Price, Is It Still a Commodity?"; Rekret, *Take This Hammer: Work, Song, Crisis*.

9. Weinbaum, *The Afterlife of Reproductive Slavery.*

10. Fortunati, *The Arcane of Reproduction.* Resisting a tendency for social reproduction feminism to be adopted as a political identity, Silvia Federici highlights that the concept of social reproduction has its origins in eighteenth-century Enlightenment-era physiocrats and economists, including François Quesnay. To this end, social reproduction does not necessitate a radical or Marxist stance. See Federici, "Social Reproduction Theory History, Issues and Present Challenges."

11. The International Wages for Housework Campaign, "The International Wages for Housework Campaign [Flyer]." Struggles over reproduction precede the 1970s: Marina Vishmidt, for example, has drawn attention to the US welfare rights movement, of which single mothers were its primary political constituent. See Vishmidt, "Human Capital or Toxic Asset."

12. Cox and Federici, *Counter-Planning from the Kitchen*, 6.

13. For an overview of social reproduction theory see Bhattacharya, *Social Reproduction Theory.*

14. For more on the tension between different concepts of reproduction see Vishmidt, "The Two Reproductions in (Feminist) Art and Theory since the 1970s."

15. For more on this point see Battistoni, "Ideology at Work?"; Endnotes, "The Logic of Gender."

16. Other trajectories have emerged in other regions. Whereas developed economies have been associated with a trajectory in which state welfare has declined, in India, for example, as with other parts of the Global South, the early 2000s saw an expansion of state subsidies for social reproduction, primarily in response to grassroots campaigns. Many of these changes were reversed, however, with the Modi regime from 2014 onwards. Rao, "Beyond the Coronavirus."

17. Cooper, *Family Values.*

18. Fraser, "Contradictions of Capital and Care."

19. Fraser, "Contradictions of Capital and Care." As I discuss in chapter 4, the UK's New Labour government (1997–2010) exemplifies how neoliberalism has not only resulted in a reduction of public spending. Rather, what is characteristic is the recomposition of welfare around market logics.

20. Benanav and Clegg, "Misery and Debt: On the Logic and History of Surplus Populations and Surplus Capital," 605. A geographic distinction can be drawn here between capital's reserve army of labor and those who experience "dispossession without proletarianization" and are largely external to capital accumulation. See Sanyal and Bhattacharya, "Beyond the Factory." Bhattacharya et al., "Exclusion, Surplus Population, and the Labour Question in Postcolonial Capitalism."

21. Fraser, "Contradictions of Capital and Care"; Adkins et al., *The Asset Economy*; Vishmidt, "Human Capital or Toxic Asset."

22. Notable exceptions include Rekret, *Take This Hammer: Work, Song, Crisis*; Drott, *Streaming Music, Streaming Capital.* Studies of music that refer to

social reproduction often draw upon the concept's use by Pierre Bourdieu. The Bourdieuian formation has some coherences but is nevertheless distinct from its use in Marxist feminism and social reproduction theory. For more on the relationship between these two intellectual lineages see Backer and Cairns, "Social Reproduction Theory Revisited."

23. Sterne, *The Audible Past*, 7.

24. Fraser, "Contradictions of Capital and Care."

25. Rao, "Beyond the Coronavirus."

26. Cammack and Gimenez, "The Permanent Global Crisis of Working-Class Social Reproduction."

27. Artificial intelligence is an imprecise designation, referring to a range of technologies that involve algorithmic, machine learning, and large language model processing.

28. Benanav, *Automation and the Future of Work.*

29. Sharma, "Going to Work in Mommy's Basement."

30. Benanav, *Automation and the Future of Work.*

31. Mueller, "The Full Automation Fallacy."

32. Gray and Suri, *Ghost Work.*

33. Battistoni, "Ideology at Work?"

34. For more on the constitution of the "high cost" care sector, see Battistoni, "Ideology at Work?"

35. Benanav and Clegg, "Misery and Debt: On the Logic and History of Surplus Populations and Surplus Capital."

36. Mueller, "The Full Automation Fallacy."

37. On music's function as a technology of surveillance, see Drott, *Streaming Music, Streaming Capital.*

38. O'Brien, *Family Abolition*, 4.

39. For example, see Hagood, *Hush*; Allen Anderson, "Neo-Muzak and the Business of Mood"; Kassabian, *Ubiquitous Listening.* An important and influential articulation of music as a technology of the self is found in DeNora, *Music in Everyday Life*, much of which prefigures the changes to music consumption with the rise of streaming.

40. Cooper, *Family Values.*

41. The now-defunct Rdio was the first streaming service to offer a family subscription in 2011. Spotify launched its family plan in 2014, purportedly in response to consumer demand. Apple Music's family plan, also launched in 2014, is part of its more general Apple family sharing feature. Tidal launched a family plan in 2015, Deezer in 2016, and Pandora and YouTube Music launched in 2018.

42. Friedman and Friedman, *Capitalism and Freedom*, 33.

43. Spotify, "Spotify Kids Is Now Available in the U.S., Canada, and France."

44. Spotify, "Introducing Spotify Kids, A New Standalone App for the Next Generation of Listeners."

45. Streaming platforms also contain other organizational logics, enabling access to recommended "new" musics, for example. However, mood-based and activity-based listenings remain a key feature of many platforms. Indeed, as discussed in chapter 5, some streaming services are specifically organized around particular activities and their affective fulfillment.

46. Taylor, "The Commodification of Music at the Dawn of the Era of 'Mechanical Music,'" 291.

47. Drott, *Streaming Music, Streaming Capital*, 208.

48. Allen Anderson, "Neo-Muzak and the Business of Mood"; Jones and Schumacher, "Muzak."

49. Critical accounts that draw connections between Muzak's history and current circulation of music in accordance with mood include Allen Anderson, "Neo-Muzak and the Business of Mood"; Rekret, "'Melodies Wander around as Ghosts.'" Lo-fi beats is a shorthand for a genre of music that has been presented as conducive for study and focus and is characterized by unobtrusive production and repetitive musical features.

50. The Muzak corporation still exists, although it is now named Mood Media, and is based in Austin, Texas.

51. Paul Allen Anderson, for example, briefly mentions Good Music, a radio format originating in the late 1950s, targeted towards women working in the home who were uninterested in youth-oriented music stations. See Allen Anderson, "Neo-Muzak and the Business of Mood."

52. Allen Anderson, "Neo-Muzak and the Business of Mood."

53. Radano, "Interpreting Muzak."

54. Jones and Schumacher, "Muzak," 161–62.

55. Sampson, "Spoti-Fail."

56. For example, see the Reddit threads "can i share my family plan with friends?," and "Spotify Family Plan, Do They Actually Check You Live There?"

## CHAPTER 1. MOTHERING MACHINES

1. "Songs My Mother Taught Me" likely refers to Antonín Dvořák's popular art song, which is a setting of Adolf Heyduk's poem of the same title or the alternative song setting of the same poem by Charles Ives, all of which have a nostalgic tenor.

2. Sousa, "The Menace of Mechanical Music," 281.

3. Warfield, "John Philip Sousa and 'The Menace of Mechanical Music.'"

4. Kraft, "Musicians in Hollywood," 294.

5. Kraft, "Musicians in Hollywood," 295.

6. Sousa, "The Menace of Mechanical Music."

7. For important analyses of how the phonograph was preceded by, converged with, and was differentiated from other sound reproduction technologies see

Gitelman, *Scripts, Grooves, and Writing Machines*; Sterne, *The Audible Past*; Taylor et al., *Music, Sound, and Technology in America*; Taylor, "The Commodification of Music at the Dawn of the Era of 'Mechanical Music.'" There were also various musical automata prior to the nineteenth century that prefigure the separation of music from human performer and establish its mechanical reproducibility. For more on this see Voskuhl, *Androids in the Enlightenment.*

8. Gitelman, *Scripts, Grooves, and Writing Machines.*

9. The lullaby is illustrative of this, as discussed in chapter 5.

10. Timothy Taylor offers an important interrogation of this narrative, highlighting how music was commodified prior to the phonograph. See Taylor, "The Commodification of Music at the Dawn of the Era of 'Mechanical Music.'"

11. Edison, "'The Woman of the Future,' Good Housekeeping (1912)," 258.

12. Edison, "'The Woman of the Future,' Good Housekeeping (1912)," 258–59.

13. Edison, "'The Woman of the Future,' Good Housekeeping (1912)," 259.

14. Edison, "'The Woman of the Future,' Good Housekeeping (1912)," 259.

15. Edison, "'The Woman of the Future,' Good Housekeeping (1912),", 266.

16. Edison, "'The Woman of the Future,' Good Housekeeping (1912)," 265–66.

17. It is notable that Edison's interview was published around the same time as Christine Frederick's *Ladies Home Journal* series, "The New Housekeeping: Efficiency Studies in Home Management," which wasp published between September and December 2012 and explained the principles of Taylorism and their applicability to the middle-class home. For more on the tensions between the "angel of the home" ideal and women's domestic experiences see McClintock, *Imperial Leather*, 160–62.

18. Mies, *Patriarchy and Accumulation on a World Scale*, 110.

19. Mies contests Sobart's characterization of the "leisure class of women" as the drivers of luxury consumption. While Sobart suggests that the courtesans were responsible for this development due to their vanity and addiction to luxurious clothing, housing, furniture, and cosmetics, Mies argues that this line of argument serves to make women the "villains": "would it not amount to saying that it was not the men—who wielded economic and political power—who were the historical 'subjects' (in the Marxist sense) but the women, as the real power behind the scenes who pulled the strings and set the tune according to which mighty men danced?" Mies, *Patriarchy and Accumulation on a World Scale*, 101.

20. Mies, *Patriarchy and Accumulation on a World Scale*, 103.

21. Mies, *Patriarchy and Accumulation on a World Scale*, 104.

22. Mies, *Patriarchy and Accumulation on a World Scale*, 104.

23. Mies, *Patriarchy and Accumulation on a World Scale*, 105.

24. In the case of the Mormons, the persecution of polygamy, the dissolution of the Mormon church's property, and opening Utah to transcontinental railroads had severe consequences for Mormons' nineteenth-century economies. Mohandesi and Teitelman, "Without Reserves," 41.

25. Mohandesi and Teitelman, "Without Reserves," 40.

26. Mies, *Patriarchy and Accumulation on a World Scale*, 106.

27. Gitelman, "How Users Define New Media."

28. Taylor et al., *Music, Sound, and Technology in America*, 20.

29. Kenney, *Recorded Music in American Life*, 90.

30. Edison, "The Phonograph and Its Future," 533.

31. Edison, "The Phonograph and Its Future." As Lisa Gitelman makes clear, Edison's proposed musical uses of the phonograph conformed with his understanding of it as a technology of dictation capable of capturing vocal utterances. See Gitelman, "Souvenir Foils."

32. Kenney, *Recorded Music in American Life*, 24.

33. Bowers, "Creating a Home Culture for the Phonograph: Women and the Rise of Sound Recordings in the United States, 1877–1913," 54.

34. Roy, "'You Ought to See My Phonograph': The Visual Wonder of Recorded Sound (1877–1900)," 191.

35. Kenney, *Recorded Music in American Life*, 45.

36. Marvin, *When Old Technologies Were New*, 6; see also Martin, *"Hello, Central?"*

37. Kruse, "Early Audio Technology and Domestic Space."

38. Marvin, *When Old Technologies Were New*, 80. In the case of the domestic telephone, Martin observes that greater opportunities for social contact for women consumers were counterbalanced with limitations on opportunities for contact outside the home: the replacement of visits to shops and friends with phone calls further confined women to domestic space. See Martin, *"Hello, Central?"*

39. Mohandesi and Teitelman, "Without Reserves," 44–45.

40. Gitelman, "How Users Define New Media."

41. Kenney, *Recorded Music in American Life*, 138–39.

42. Hyman, "Immigrant Women and Consumer Protest"; Mohandesi and Teitelman, "Without Reserves."

43. *Success Magazine & the National Post*, 62.

44. Taylor et al., *Music, Sound, and Technology in America* 4.

45. Rekret, *Take This Hammer: Work, Song, Crisis*, 27–28.

46. Kenney, *Recorded Music in American Life*, 90.

47. On this point see also Gitelman, "How Users Define New Media."

48. Solie, *Music in Other Words*, 91.

49. Vorachek, "The Instrument of the Century," 26.

50. Vorachek, "The Instrument of the Century"; Solie, *Music in Other Words*.

51. Miller, "Working Musicians," 428.

52. Miller, "Working Musicians."

53. Solie, *Music in Other Words*, 95.

54. Solie, *Music in Other Words*, 95.

55. In this regard, it was framed in similar ways to the player piano: a contemporaneous yet initially more popular technology that lay many of the foundations for domestic phonography. See Taylor, "The Commodification of Music at the Dawn of the Era of 'Mechanical Music.'"

56. Katz, *Capturing Sound*, 58–59.

57. On this point, see also Laing, "A Voice without a Face," 5.

58. *The Ladies Home Journal*, "The Amberola: A New Edison Phonograph," 24.

59. For an illustrative discussion of how recording separates music from function in capitalist modernity see Denning, *Noise Uprising*, 8.

60. Hui, "Lost."

61. Hui, "Lost."

62. Kenney, *Recorded Music in American Life*, 10–11.

63. Kenney, *Recorded Music in American Life*, 11.

64. Kenney, *Recorded Music in American Life*, 11.

65. As Kenney notes, many of Victor's opera recordings were in fact popular and folk songs sung in an operatic style by opera stars. Indeed, Kenney argues that "to an important degree, Victor imposed its own 'high class' gloss on opera, which was not especially considered high class in Italy, for example, when compared to symphonic or chamber music." Kenney, *Recorded Music in American Life*, 50.

66. A copy of this advertisement can be found in Taylor et al., *Music, Sound, and Technology in America*, 70.

67. "Music in the Home."

68. Bowers, "Creating a Home Culture for the Phonograph: Women and the Rise of Sound Recordings in the United States, 1877–1913," 182.

69. Bowers, "Creating a Home Culture for the Phonograph: Women and the Rise of Sound Recordings in the United States, 1877–1913," 182.

70. On this point see also Sterne, *The Audible Past*.

71. Thompson, "Time, Work-Discipline, and Industrial Capitalism." I return to clock-time and the separation of work and rest in chapter 5.

72. Bailey, *Leisure and Class in Victorian England*, 4.

73. Bailey, *Leisure and Class in Victorian England*, 4.

74. Bailey, *Leisure and Class in Victorian England*, 102.

75. Bailey, *Leisure and Class in Victorian England*, 6.

76. Weeks, *The Problem with Work*, 14.

77. McClintock, *Imperial Leather*, 162.

78. Cowan, *More Work for Mother*, 42. The divisions of leisure and nonleisure, obligation and choice are further destabilized by the historical expectations placed upon women's recreational activities. Women of all classes have been implored to make their leisure time "useful" for capitalist society, in alignment with bourgeois and heteronormative ideals of reproductive respectability. This principle is well illustrated by late nineteenth-century philanthropists and

reformers who sought to shape working-class women's leisure time to encourage self-improvement. Catriona M. Parratt notes how women reformers in late Victorian and Edwardian England were anxious about the abilities of working-class families to raise young women and girls with the skills to be good wives and mothers, those makers of healthy and happy homes that nation and empire needed. Affluent women reformers understood themselves as able to cross the social and cultural divide separating them from their working-class counterparts and, through personal influence and example, sought to direct their recreational activities. This "critical and difficult work" was claimed by reformers as their own "work for which they were singularly fitted." Parratt, "Making Leisure Work," 477.

79. For an important historical overview of women's role in establishing a domestic telephone culture see Martin, *"Hello, Central?"*

80. Fischer, "Gender and the Residential Telephone," 227.

81. Marx and Engels, *The Communist Manifesto*, 79–80.

82. This point echoes many discussions about "free" labor within digital culture, some of which will be addressed in chapter 2.

83. Marx, *Capital*, 132.

84. Fortunati, *The Arcane of Reproduction*, 8.

85. Fortunati, *The Arcane of Reproduction*, 31.

86. Endnotes, "The Logic of Gender."

87. Stakemeier and Vishmidt, *Reproducing Autonomy*.

88. Gonzalez, "The Gendered Circuit."

89. Dalla Costa and James, "Women and the Subversion of the Community," 44.

90. Dalla Costa and James, "Women and the Subversion of the Community," 45.

91. Dalla Costa and James, "Women and the Subversion of the Community," 45.

92. Cowan, *More Work for Mother*.

93. Dalla Costa and James, "Women and the Subversion of the Community," 29. This notion is also illustrated by the Endnotes collective apropos of childcare: where some activities concerning childcare might be performed more quickly, they still need to be looked after for the whole day and that amount of time is not flexible. See Endnotes, "The Logic of Gender."

94. Endnotes, "The Logic of Gender."

95. Edison, "'The Woman of the Future,' Good Housekeeping (1912)," 261.

96. Atanasoski and Vora, *Surrogate Humanity*, 3–4.

97. Mies, *Patriarchy and Accumulation on a World Scale*, 74.

98. Mies, *Patriarchy and Accumulation on a World Scale*, 77.

99. Mies, *Patriarchy and Accumulation on a World Scale*, 103.

100. Laing, "A Voice without a Face"; Weheliye, *Phonographies*.

101. Weheliye, *Phonographies*, 39.

102. Weheliye, *Phonographies*, 40.

103. Gitelman, "Reading Music, Reading Records, Reading Race," 267.

104. Taussig, *Mimesis and Alterity*.

105. Marvin, *When Old Technologies Were New*, 39.
106. See Devine, *Decomposed*, 55–58.
107. Kenney, *Recorded Music in American Life*, 7.
108. Kheshti, *Modernity's Ear*, 28.
109. Kheshti, *Modernity's Ear*, 23.
110. Kheshti, *Modernity's Ear*, 4.
111. Kheshti, *Modernity's Ear*, 132.
112. Denning, *Noise Uprising*, 113.
113. Denning, *Noise Uprising*, 140.

## CHAPTER 2. "HEAR FOR YOU"

1. "Amazon Alexa: Sharing is caring."
2. Edison Research, "UK Smart Speaker Ownership Outpaces U.S."
3. Koksal, "The Sales of Smart Speakers Skyrocketed."
4. Bohn, "Amazon Says 100 Million Alexa Devices Have Been Sold."
5. "Alexa for Business."
6. Strengers and Kennedy, *The Smart Wife*.
7. *Amazon Echo: Remember Baby*.
8. Phan, "Amazon Echo and the Aesthetics of Whiteness."
9. In England, for example, the NHS reports in 2019 that over 80% of adult social care roles were undertaken by women. While the 2019 workforce is predominantly white, when standardized per 100,000 of the population, it becomes clear that a higher proportion of African, Caribbean, and Black British workers are involved in social care by comparison to white workers. England's independent and local authority Adult Social Care workforce (i.e., the workforce outside the NHS) is also predominantly female: a 2019 report found that 83% of roles are undertaken by women; around 21% of workers identified as Black, Asian, or Minority Ethnic (BAME); and 11% identified as Black, who make up 3% of the total population. See Skills for Care, *The State of the Adult Social Care Sector and Workforce in England*; NHS England Digital, "Individual Worker Characteristics."
10. Roberts, "Echo and the Chorus of Female Machines."
11. Phan, "Amazon Echo and the Aesthetics of Whiteness."
12. "Divorced Siri," my emphasis.
13. Voegelin, *Sonic Possible Worlds*, 167. My emphasis.
14. LaBelle, *Sonic Agency*, 145.
15. For more on feminist theories of listening see Farinati and Firth, *The Force of Listening*.
16. Bourgault, "Attentive Listening and Care in a Neoliberal Era," 331. Bourgault's claim that online distance learning has correlated with a decline in listening skills remains limited in its substantiation: there is no attention, for example,

to different modes and models of online distance learning nor to the different topics that may be examined through online distance learning.

17. Grossman, "The Feminised Service Sector."

18. Bridget Anderson's study of the global politics of domestic labor, for example, highlights how racial, cultural, and religious hierarchies inform and are reproduced by the organization of paid domestic labor in Europe, feeding into who is deemed more or less appropriate for undertaking different types of domestic work with different levels of wages attached. Anderson also elucidates how when sourcing workers, employers moved "from nationality to nationality until they found one that suited them" suggesting that this constitutes "an extension of the naturalization of domestic work—to find a good worker one has simply to find the right genes." Anderson, *Doing the Dirty Work?*, 155.

19. As Aronson and Neysmith make clear, this situation leaves home care workers open to further exploitation. One interviewee, for example, explains: they "sometimes stay extra at night and sit and talk or, if they're watching a soap opera—which I'm not into—but because they're watching it and they tell me how good it is, I'll sit with them for half an hour or so." Aronson and Neysmith, "You're Not Just in There to Do the Work," 68.

20. Rollins, *Between Women*, 166.

21. Rollins, *Between Women*, 167.

22. Katzman, *Seven Days a Week*, 188.

23. Rollins, *Between Women*, 209.

24. Jin and Wang, Voice-based determination of physical and emotional characteristics of users.

25. Jin and Wang, Voice-based determination of physical and emotional characteristics of users.

26. Andrejevic, *Infoglut*, 43–44.

27. Some examples of texts on digital media, technologies, and labor that make reference to Hochschild's *The Managed Heart* include Arcy, "Emotion Work"; and Duffy, "The Romance of Work."

28. Listening's relation to emotional labor is largely absent in Hochschild's analysis: it is gestured towards in references to conversation but remains untheorized as a means of managing feeling. In one of the few overt references to listening work, Hochschild notes the ways in which flight attendant roles are differently gendered, inasmuch as women workers are asked more often "to appreciate jokes and listen to stories." Hochschild, *The Managed Heart*, 177.

29. Hochschild, *The Managed Heart*, 20.

30. Hochschild, *The Managed Heart*.

31. Hochschild, *The Managed Heart*, 132.

32. Hochschild, *The Managed Heart*, 90.

33. Hanley and Azenkot, "Understanding the Use of Voice Assistants by Older Adults."

34. Schiller and McMahon, "Alexa, Alert Me When the Revolution Comes."

35. Weeks, "Life within and against Work: Affective Labour, Feminist Critique, and Post-Fordist Politics," 244.

36. Whitney, "Affective Indigestion: Lorde, Fanon, and Gutierrez-Rodriguez on Race and Affective Labor," 283.

37. Gutiérrez-Rodríguez, *Migration, Domestic Work and Affect.*

38. Dreyfuss, "The Terrible Joy of Yelling at an Amazon Echo."

39. Larson, "Yelling at Amazon's Alexa."

40. Andrews, "Alexa, Just Shut Up."

41. Dreyfuss, "The Terrible Joy of Yelling at an Amazon Echo."

42. Haraway, "A Cyborg Manifesto: Science, Technology, and Socialist-Feminism in the Late Twentieth Century," 153.

43. For examples of this interpretation see Zylinska, *The Cyborg Experiments.* This reading of Haraway is also prevalent within computer music research, where the use of wearable tech in performance is often articulated with reference to the cyborg. See for example Hsu and Kemper, "Enacting Sonic-Cyborg Performance through the Hybrid Body in Teka-Mori and Why Should Our Bodies End at the Skin?"

44. For examples of this tendency see Phan, "Amazon Echo and the Aesthetics of Whiteness"; Strengers and Kennedy, *The Smart Wife.*

45. Crawford and Joler, *Anatomy of an AI System.* See also "Anatomy of an AI System," https://anatomyof.ai/.

46. For more on the relationship between digital culture, resource extraction, and the geopolitics of labor see Miller, "The Art of Waste: Contemporary Culture and Unsustainable Energy Use"; Taffel, "Escaping Attention: Digital Media Hardware, Materiality and Ecological Cost."

47. Chan et al., *Dying for an iPhone.*

48. Chan et al., *Dying for an iPhone.*

49. Day et al., "Thousands of Amazon Workers Listen to Alexa Conversations"; Adolph and McNicols, "Meet the Coders Who Hear You Yell at Alexa."

50. Day et al., "Thousands of Amazon Workers Listen to Alexa Conversations."

51. During the development of Echo and Alexa, the generation of training datasets from recorded user interactions, which could help the device respond appropriately to a wide range of commands and voices within different auditory environments, was reportedly staged using a large temporary workforce. Acting on behalf of Amazon, the Australian data firm Appen rented properties in various US cities, and hid multiple Echo prototypes within them. Contract workers would be asked to attend the property and read a script and respond to various written cues. See Stone, *Amazon Unbound.*

52. Crawford, *Atlas of AI.*

53. Terranova, "Free Labor."

54. For a theorization of attention as labor, see Bueno, *The Attention Economy.*

55. Drott, *Streaming Music, Streaming Capital.*

56. Kelion, "Amazon: How Bezos Built His Data Machine."

57. Srnicek, *Platform Capitalism*; Christophers, *Rentier Capitalism.*

58. Chan, "Amazon Web Services Generated over $10 Billion in Quarterly Revenue."

59. Amazon, "Help: Privacy Notice."

60. Amazon, "Help: Interest-Based Ads."

61. Amazon, "Advertising:Amazon DSP (Demand Side Platform)."

62. Leber, "Amazon Woos Advertisers with What It Knows about Consumers."

63. Kaplan, "The Self-Consuming Commodity."

64. See for example Hesmondhalgh, "User-Generated Content, Free Labour and the Cultural Industries."

65. For example, see Terranova, "Free Labor," 54.

66. Crawford and Joler, *Anatomy of an AI System.* See also Dyer-Witheford, *Cyber-Proletariat.*

67. Dyer-Witheford, *Cyber-Proletariat.*

68. For a summary of Fuchs's thesis on value production and free digital labor, see Fuchs, "The Digital Labour Theory of Value and Karl Marx in the Age of Facebook, YouTube, Twitter, and Weibo."

69. Beech, *Art and Postcapitalism*, 97.

70. Beech, *Art and Postcapitalism*, 96.

71. Indeed, the smart speaker is symptomatic of the coherence between the literal forms of extraction involved in the mining industries and more recent forms of data extraction facilitated by digital technologies. For an insightful analysis of the relationships between these different forms of capitalist extraction, see Mezzadra and Neilson, "On the Multiple Frontiers of Extraction."

72. Robinson, "Databases and Doppelgängers," 419.

73. Wright, "Amazon to Pay $25m over Child Privacy Violations."

74. Lynskey, "Alexa, Are You Invading My Privacy?"

75. McCarthy, "LAUREN Testimonials."

76. McCarthy, "Feeling at Home: Between Human and AI."

77. Capgemini, "Smart Talk: How Organizations and Consumers Are Embracing Voice and Chat Assistants."

78. Manikonda et al., "What's up with Privacy?"

79. "Best Super Affordable Alexa Devices for 2020."

80. *Online Platforms and Market Power Part 6: Hearing before the Subcommittee on Antitrust, Commercial and Administrative Law,* 116th cong, 116-94 (2020).

81. Stone, *Amazon Unbound: Jeff Bezos and the Invention of a Global Empire.*

82. Woods, "Asking More of Siri and Alexa," 335–36.

83. Woods, "Asking More of Siri and Alexa," 346.

84. Moynihan and Schwär, "Amazon May Have given Alexa Female Voice to Make Device Sound 'Caring.'"

85. For an important critique of such framings of oral/aural mediation see Sterne, "The Theology of Sound."

86. Alexa Skills Kit, "What Is Conversational AI?"

87. Stone, "The Secret Origins of Amazon's Alexa."

88. Stone, *Amazon Unbound: Jeff Bezos and the Invention of a Global Empire.* According to Stone, the product team in charge of marketing and selling the device instead emphasized features categorized as "magical and mundane": basic and reliable services such as setting alarms and checking weather reports. It is this vision for the technology that is reflected in the "Sharing is caring" advert.

89. Dredge, "Everybody's Talkin': Smart Speakers and Their Impact on Music Consumption."

90. Rekret, *Take This Hammer: Work, Song, Crisis.*

91. For an important and influential examination of music as a technology of the self which precedes streaming services see DeNora, *Music in Everyday Life.*

92. Drott, *Streaming Music, Streaming Capital*, 31.

93. West, "Amazon," 31.

94. Wright, *Technology in Social Care.*

95. Watkins et al., "Effects of Health and Social Care Spending Constraints on Mortality in England."

96. Local Government Association, "Hampshire County Council: Pushing the Boundaries by Using Amazon Echo."

97. PA Argenti, "Insights."

98. Wright, *Technology in Social Care*; Learner, "Over Half of Staff Think Care Homes Should Use Alexa to Help Care for Residents."

99. *BBC News*, "Why Is a Council Giving Out Virtual Assistants?"

100. Amazon News, "Alexa Can Now Help Make the Remote Caregiving Experience Easier."

101. Local Government Association, "Hampshire County Council: Pushing the Boundaries by Using Amazon Echo."

102. Oxfordshire City Council, "Care Has Just Got a Lot Smarter."

103. Oxfordshire City Council, "Care Has Just Got a Lot Smarter."

104. Community Care, "Talking Alexa: How Assistive Technology Can Help Adult Social Care Services."

105. While techno-austerity draws attention to the ways that technological interventions are presented as solutions to the demands of austerity, the notion that technology can cut the costs of care precedes the 2010 implementation of austerity in the UK: a 2004 Audit Commission report, for example, states that assistive technologies offer "the tantalising possibility for public policy to meet more people's desire to remain independent for longer, while at the same time saving money overall." See Wright, *Technology in Social Care.*

106. The pervasiveness of this ideological formation was highlighted in 2019 by the shadow cabinet minister Jo Platt, who warned in a speech to the Labour Party conference that "technology should not be a codeword for austerity," with

narratives of "efficiency" and "technological innovation" having been used to mask cuts to public services and the welfare system. Politics Home, "Technology Can Improve Public Services but It Must Not Be a 'Codeword for Austerity,' Warns Shadow Cabinet Office Minister."

107. Government and Social Care Ombudsman, "Devon County Council (17 010 182)."

108. The notion of "just ask" also misses the in-built racial disparities of voice-recognition technologies: African American users, for example, have been found to be disproportionately misheard by Alexa and other automated speech recognition systems by comparison to white users. See Koenecke et al., "Racial Disparities in Automated Speech Recognition."

109. As Jennifer Lynch et al. argue, public health rhetoric around technologically facilitated independence requires critical interrogation: aspirations for "independence," "staying at home," or "achieving self-management" are by no means benign descriptors, nor do they necessarily mean the same thing for care providers and service users. See Lynch et al., "If Telecare Is the Answer, What Was the Question?"

110. Johnston and Pratt, "Automating Adult Social Care in the UK."

111. Wright, *Technology in Social Care*; Johnston and Pratt, "Automating Adult Social Care in the UK."

112. Johnston and Pratt, "Automating Adult Social Care in the UK."

## CHAPTER 3. "YOUR WOMB, THE PERFECT CLASSROOM"

1. An earlier, shorter version of this chapter was published as the following article: Thompson, "'Your Womb, The Perfect Classroom.'"

2. Institut Marquès, *Soraya Arnelas' Christmas Concert for Future Babies via the Babypod®*.

3. Institut Marquès, "Music and Fertilisation."

4. Babypod, "What Is Babypod?"

5. See W. Shannon Thomas Jr.'s Patent US4934998 for prenatal audio apparatus, filed on 10 July 1986.

6. BabyPlus, "BabyPlus® Prenatal Education System."

7. BabyPlus, "About."

8. The BabyPlus Company USA, *Prenatal Education*.

9. Institut Marquès, "Music and Fertilisation."

10. Thompson, *Making Parents*, 8.

11. Foucault, *The Birth of Biopolitics*.

12. Foucault, *The Birth of Biopolitics*, 229.

13. Brown, *Undoing the Demos*, 106.

14. Foucault, *The Birth of Biopolitics*, 230.

15. Murphy, "The Girl: Mergers of Feminism and Finance in Neoliberal Times."

16. Schultz, "Investment in Human Capital," 3.

17. Becker, *A Treatise on the Family*, 222; Cooper, *Family Values*, 225.

18. Heckman and Masterov, "The Productivity Argument for Investing in Young Children," 446–47.

19. Heckman and Masterov, "The Productivity Argument for Investing in Young Children," 487.

20. Heckman's own original version of the curve does not include the prenatal bar and instead shows ages 0–3 as the referent for interventions targeting the earliest years.

21. Notably, Heckman, like Shultz, has argued that investment should be both private *and* public, insofar as childhood education initiatives offer both individual and societal returns. See Heckman, "Promoting Social Mobility."

22. For example, see Hallam and Music Education Council, *The Power of Music*. See also Purves, "Local Authority Instrumental Music Tuition as a Form of Neo-Liberal Parental Investment."

23. Ilari, "Concerted Cultivation and Music Learning."

24. BabyPlus, "About."

25. Lareau, *Unequal Childhoods*. See also Vincent, "'The Children Have Only Got One Education and You Have to Make Sure It's a Good One.'"

26. BabyPlus, "About."

27. BabyPlus, "FAQs." No citations are provided for the studies referenced.

28. Institut Marquès, *Which Music Stimulates Your Baby?*

29. Mammoser, "Meet the Doctor behind the World's First Vaginal Speaker."

30. Fancourt and Perkins, "Could Listening to Music during Pregnancy Be Protective against Postnatal Depression and Poor Wellbeing Post Birth?"

31. For an important sociological account tracing the cultural and political emergence and implications of the Mozart effect see Beauvais, "The 'Mozart Effect.'"

32. Babypod, "What Music Do Foetuses Prefer?"

33. Mammoser, "Meet the Doctor behind the World's First Vaginal Speaker."

34. Mozart's K488 is incorrectly listed as a piano sonata.

35. Rekret, *Down with Childhood*, 78–79.

36. Katz, "Childhood as Spectacle: Relays of Anxiety and the Reconfiguration of the Child," 6.

37. Katz, "Childhood as Spectacle: Relays of Anxiety and the Reconfiguration of the Child," 11.

38. Katz, "Childhood as Spectacle: Relays of Anxiety and the Reconfiguration of the Child"; Katz, "Accumulation, Excess, Childhood."

39. My intention here is not to suggest that this entry is "new": the occupation of the reproductive body by capitalism may be seen, for example, within the

histories of slavery and colonialism, whereby pregnancy and childrearing has been either enforced or prohibited, depending on its profitability. For more on this see Mies, *Patriarchy and Accumulation on a World Scale*; Weinbaum, *Wayward Reproductions.*

40. The Pregnancy Zone, "9 Best Prenatal Sounds and Systems for Baby and Mom."

41. Lords Pharmacy, "SneakPeek Heartbeat Monitor."

42. Baby Doppler, "Sonoline B Fetal Doppler."

43. Lentjes, "Sounds of Life: Fetal Heartbeat Bills and the Politics of Animacy."

44. Hutton, "'Dangerous' Home Dopplers Should Be Banned."

45. Agar, "Liberal Eugenics"; Sparrow, "A Not-So-New Eugenics: Harris and Savulescu on Human Enhancement."

46. Agar, "Liberal Eugenics"; *Harvard Law Review*, "Regulating Eugenics."

47. For more on this see Daar, *The New Eugenics*; Daniels and Heidt-Forsythe, "Gendered Eugenics and the Problematic of Free Market Reproductive Technologies."

48. Francis, *Inquiries into Human Faculty and Its Development.*

49. Stanton, "Report on Use of Seashore Tests at Eastman School of Music."

50. Cowan, "Unsound: A Cultural History of Music and Eugenics."

51. Baynton, *Forbidden Signs.*

52. Sterne, *The Audible Past*, 40.

53. Bell's position on the prohibition of deaf intermarriage seemingly shifted between the 1880s and 1900. His *Memoir Upon the Formation of a Deaf Variety of the Human Race*, first presented in 1883 to the National Academy of the Sciences, suggested that legislative interventions into deaf intermarriage would be ill advised. Instead, he advocated for the determination and removal of the underlying causes that promoted deaf intermarriage. As the eugenics movement became more punitive in its approach, however, Bell sought to further discourage legislative prohibitions into deaf intermarriage. Bell's resistance to legislative prohibitions, amongst other factors, have led some to challenge the notion that his views on the Deaf community can be accurately described as eugenicist. For more on this see Greenwald and Van Cleve, "A Deaf Variety of the Human Race: Historical Memory, Alexander Graham Bell, and Eugenics."

54. Logan, *Learning before Birth*, 122.

55. BabyPlus, "About."

56. Schuller, *The Biopolitics of Feeling*, 21–22.

57. Weingarten, "From Maternal Impressions to Eugenics: Pregnancy and Inheritance in the Nineteenth-Century U.S.," 303–5.

58. Shelly, "The Maternal Impression Superstition, and Why It Survives," 865.

59. Melendy, *The Laws of Nature Revealed*, 310.

60. Melendy, *The Laws of Nature Revealed*, 295.

61. Richardson, *The Maternal Imprint*, 165.
62. Richardson, *The Maternal Imprint.*
63. Keaney, "The Racializing Womb."
64. Richardson, *The Maternal Imprint.*
65. Valdez, "The Redistribution of Reproductive Responsibility," 438.
66. Logan, *Learning before Birth*, 126.
67. Logan, *Learning before Birth*, 124.
68. Logan, *Learning before Birth*, 129.
69. The BabyPlus for Good foundation (www.babyplusforgood.org) does describe itself as promoting research initiatives into BabyPlus's benefits for children who will be born into "disadvantaged" circumstances, while one of the aims of the foundation is to "initiate and participate in research partnerships to explore the positive impact of early sound enrichment on the incidence and severity of autism/ASD in children."
70. Russell, "The Race Idea in Reproductive Technologies."
71. Reynolds, *Energy Flash*, 172.
72. Nancy, *Listening*; Sloterdijk, *Bubbles.*
73. Schafer, *The Soundscape*, 15.
74. Schafer, *The Soundscape*, 18.
75. Schafer, *The Soundscape*, 18.
76. Pettman, *Sonic Intimacy*, 1.
77. Murch, "FILM; Stretching Sound to Help the Mind See."
78. Sonic Womb Productions, "The Sonic Womb Orrb Listening Experience at the Brain Forum 2016."
79. Sonic Womb Productions, "The Sonic Womb Orrb Listening Experience at the Brain Forum 2016."
80. Sonic Womb Productions, "The Sonic Womb Orrb Listening Experience at the Brain Forum 2016."
81. Lagercrantz and Changeux, "The Emergence of Human Consciousness."
82. Stabile, "Shooting the Mother," 180.
83. Stabile, "Shooting the Mother," 180.
84. Lewis, "Amniotechnics."
85. The US continues to have the highest rate of maternal deaths of any high-income country, most of which affect Black women and are likely preventable. See Gunja et al., *Insights into the U.S. Maternal Mortality Crisis.*
86. Goh, "Sounding Situated Knowledges."
87. Logan, *Learning before Birth*, 118–19.
88. Logan, *Learning before Birth*, 3.
89. BabyPlus, "Supporting Science."
90. Babypod, "What Is Babypod?'"
91. Weheliye, "'Feenin,'" 30.
92. Evens, *Sound Ideas*, 7. My emphasis.

93. For a feminist reading of mediums of containment see Rodgers, *Pink Noises*; Sofia, "Container Technologies."

94. See Lewis, *Full Surrogacy Now.*

95. Stabile, "Shooting the Mother," 199.

96. Lewis, *Full Surrogacy Now.*

97. McLeod, "Renovate My Uterus: Aging Queerly through Performance Art and Karaoke."

98. McLeod, "Renovate My Uterus: Aging Queerly through Performance Art and Karaoke."

99. McLeod, "What a Feeling."

100. Chapter 1's discussion of Maria Mies's housewifization offers a clear illustration as to how capital understands pregnancy as tied to the renewal of capitalism's workforce.

## CHAPTER 4. AUDITORY REVANCHISM, ANTISOCIALITY, AND REPRODUCTIVE RESPECTABILITY

1. Mcardle and Erzen, *Zero Tolerance*, 42.

2. Mcardle and Erzen, *Zero Tolerance.*

3. Smith, *The New Urban Frontier*, 207.

4. Smith, *The New Urban Frontier*, 220.

5. Smith, "Giuliani Time," 1.

6. Smith, "'Giuliani Time," 3.

7. Smith, "Giuliani Time," 3.

8. Marx, *Capital*, 784.

9. Benanav and Clegg, "Misery and Debt."

10. Smith, *The New Urban Frontier.*

11. Radovac, "The 'War on Noise,'" 757.

12. Alli, "Carceral Educations."

13. Hong, *Death beyond Disavowal: The Impossible Politics of Difference*, 20.

14. DeVerteuil, "Post-Revanchist Cities?"

15. Smith, *The New Urban Frontier*; Wang, *Carceral Capitalism.*

16. Gilmore Wilson and Gilmore, "Beyond Bratton."

17. Writing in the context of New Labour's Family Interventions Programme, Caroline Hunter and Judy Nixon note how both popular discourse and legislative judgements regarding antisocial behavior have held women accountable for failing to control their families' behavior. See Hunter and Nixon, "Disciplining Women."

18. *BBC News*, "Wrexham McDonald's Plays Classical Music to Deter Bad Behaviour."

19. Pertinent to chapter 1, it is notable that classical music offered a means of bolstering the middle-class respectability of the phonograph as a "serious instrument rather than public novelty or 'toy.'" See Ehrlich, *The Piano*, 185.

20. Jackson, "Music to Deter Yobs By."

21. Hirsch, *Music in American Crime Prevention and Punishment.*

22. *BBC News*, "Symphonies Soothe Train Vandals."

23. Jackson, "Music to Deter Yobs By."

24. Duchen, "Mind the Bach."

25. English Heritage, *Heritage Crime Prevention.*"

26. Northern Media Centre, "Northern to Use Classical Music to 'Handel' Anti-Social Behaviour at Stations."

27. Newburn and Jones, "Symbolizing Crime Control."

28. Newburn and Jones, "Symbolizing Crime Control," 228.

29. Newburn and Jones, "Symbolizing Crime Control," 228.

30. For more on the specificities of New Labour's neoliberalism see Gilbert, "The Second Wave: The Specificity of New Labour Neo-Liberalism."

31. Garrett, "Making 'Anti-Social Behaviour.'"

32. Hall et al., *Policing the Crisis*, 220.

33. While New Labour emphasized the ubiquity of antisocial behavior and its impact on both urban and rural communities, as Andrew Millie has highlighted, concerns about antisocial behavior typically related to urban environments. See Millie, "Anti-Social Behaviour in British Cities."

34. For more on this, and the ways that antisocial behavior policing affected different social groups, see Squires, *ASBO Nation*.

35. Smith, *The New Urban Frontier*, 222.

36. Smith, *The New Urban Frontier*, 209.

37. Smith, "Giuliani Time," 3.

38. Anti-Social Behaviour Act 2003, § 29.

39. For more on the politics of the loiterer see Buck-Morss, "The Flaneur, the Sandwichman and the Whore."

40. Jackson, "Music to Deter Yobs By."

41. Jackson, "Music to Deter Yobs By."

42. For an earlier discussion of this distinction and its connection to revanchism, see Thompson, "To Soothe or Remove?"

43. *Sky News*, "Big Bach and Fries: McDonald's Cuts Anti-Social Behaviour with Classical Music."

44. Bull, *Class, Control, and Classical Music*, 14.

45. Bull, *Class, Control, and Classical Music.*

46. Hess, "Interrupting the Symphony," 15.

47. Among the most famous of these programs is El Sistema, which began in Venezuela and now has music education schemes in various countries, including the UK. For an important overview and critique of this program see Baker, *El Sistema.*

48. *BBC News*, "Symphonies Soothe Train Vandals."

49. Duchen, "Mind the Bach."

50. Hirsch, *Music in American Crime Prevention and Punishment*, 28.

51. Gioia, "Bach at the Burger King."

52. Bull, *Class, Control, and Classical Music.*

53. Ritchey, "Musical Development."

54. Blair himself was an MP for Sedgefield in County Durham from 1983 to 2007, while many other key figures from Blair's government held seats in the North East.

55. Miles, "'Our Tyne.'"

56. Armstrong, "Creating Sustainable Communities in 'NewcastleGateshead,'" 282.

57. Armstrong, "Creating Sustainable Communities in 'NewcastleGateshead,'" 281.

58. Villarreal, "Geographies of Social Justice." More recently, Gateshead council has been selling off aspects of its property portfolio close to the Quayside to private investors, one effect of which has been the displacement of a DIY and grassroots community music space. See Rezaei and Denton, "The Old Police House (TOPH)."

59. Lees et al., *Planetary Gentrification.*

60. Alves, "Our Fire This Time! The Favela As Zone Of Death And Insurgent Spatiality."

61. Garmany and Richmond, "Hygienisation, Gentrification, and Urban Displacement in Brazil," 128.

62. Garmany and Richmond, "Hygienisation, Gentrification, and Urban Displacement in Brazil," 128–29.

63. Alves, "Our Fire This Time! The Favela As Zone Of Death And Insurgent Spatiality."

For more on São Paulo's segregated development and its relationship to fear of crime, see Caldeira, *City of Walls.*

64. Alves, "Our Fire This Time! The Favela As Zone Of Death And Insurgent Spatiality."

65. Alves, *The Anti-Black City,* 128.

66. Alves, *The Anti-Black City.* 78.

67. Alves, *The Anti-Black City,* 130.

68. Smith, "Global Social Cleansing," 73.

69. Alves, *The Anti-Black City,* 82.

70. Vieira de Oliveira, "Weaponizing Quietness."

71. See Lippman, "Law for Whom?" As Lippman notes, *funk* has been subject to "culturalization" as well as "criminalization." Some of the *funk* communities have responded to the repression of *baile funk* through engaging with legal and bureaucratic channels and appealing to liberal notions of racism and classism in order to assert *funk* as a legitimate culture.

72. Cardoso, *Sound-Politics in São Paulo,* 177.

73. Bonnisseau, "Ostentação."

74. Bonnisseau, "Ostentação."

75. Barbara, "Whose Mall Is It?"

76. Cardoso, *Sound-Politics in São Paulo.*

77. Alves, *The Anti-Black City*, 92.

78. Alves, *The Anti-Black City*, 86.

79. Alves, *The Anti-Black City*, 87.

80. Souza and Sá, "Em São Paulo, Operação 'Pancadão' Reprime Bailes Na Periferia." The original reads: "Para a 'sociedade,' a polícia está apenas reprimindo o crime . . . garantindo a paz e o sossego aos moradores 'trabalhadores' para que possam cumprir seus expedientes no dia seguinte."

81. These expressed anxieties about teenagers becoming pregnant at pancadãoes are reported in both Alves's and Cardoso's ethnographic accounts of community security councils.

82. Vieira de Oliveira, "Weaponizing Quietness."

83. Vieira de Oliveira, "Weaponizing Quietness."

84. The translation of this remark appears in Vieira de Oliveira, "Weaponizing Quietness," 199.

85. Cardoso, *Sound-Politics in São Paulo*, 15.

86. Vieira de Oliveira, "Weaponizing Quietness," 201.

87. Vieira de Oliveira, "Weaponizing Quietness," 203.

88. McLennan, "New Plan to Tackle Anti-Social Behaviour Based on Darwin."

89. Humbugging is an imprecise term with a variety of connotations but is used in this context to refer to behavior that involves pestering or irritating someone.

90. Moreton-Robinson, *The White Possessive.* An illustrative selection of critical social media responses to the app can be found in an article by the Daily Mail Online: Mazzoni, "Government Is Accused of 'criminalising Homelessness' after App Asks People to Dob in Rough Sleepers."

91. Moreton-Robinson, *The White Possessive*, 161.

92. Howse, "Towards a Dealing Just and Kind."

93. Howse, "Towards a Dealing Just and Kind."

94. Tamara and Taylor, "'You're Not Welcome Here': Police Move-On Powers and Discrimination Law."

95. Moreton-Robinson, *The White Possessive*, 155.

96. Haebich, *Broken Circles.*

97. These population figures are reported on the Northern Territory Government website (see Northern Territory Government, "Population." In 2012, the city was recorded as having the most expensive housing of any Australian capital other than Sydney. However, following a peak in 2014 and up into late 2019, housing prices were in a period of relative decline. More recently, housing prices in Darwin have been increasing once again. See Northern Territory Government, "Housing"; Pawson et al., *Australian Homelessness Monitor 2018.*

98. Pawson et al., *Australian Homelessness Monitor 2018.* The report takes a broad approach to defining homelessness. It primarily focuses on rough sleeping but also recognizes that homelessness affects the broader population as insecure,

overcrowded, or otherwise unsatisfactory housing. The report cites the Australian Board of Statistics definition of homelessness as referring to a person who is entirely roofless or occupies a dwelling that is physically inadequate, provides no or only a short and nonextendable tenure, or enables the resident no control of or access to space for social relation.

99. Parsell and Phillips, "Indigenous Rough Sleeping in Darwin, Australia."

100. Fisher, "Becoming the State in Northern Australia."

101. Pollard et al., "Contested Spaces"; Pollard, "Archaeology in the Long Grass: A Study of Aboriginal Fringe Camps in Darwin, Australia."

102. Fisher, "Running Amok or Just Sleeping Rough?," 171.

103. Fisher, "Becoming the State in Northern Australia."

104. Pollard et al., "Contested Spaces." Such practices also raise tensions between Indigenous and non-Indigenous constructions of public and private space: as Eric Michaels highlights, in traditional communities in remote areas much of life is conducted outdoors. These spaces, however, are not necessarily considered public in ways that cohere with settler understandings of public space. See Michaels, *Bad Aboriginal Art*, 63.

105. Thompson, *The Soundscape of Modernity*, 124.

106. Thompson, *The Soundscape of Modernity*, 124–25.

107. Clifford Rosenberg, "Shhh! Noisy Cities, Anti-Noise Groups and Neoliberal Citizenship."

108. Herring, "Complaint-Oriented Policing," 797.

109. Larrakia Nation, "The Larrakia People." The formal recognition of Larrakia ownership, via the Kenbi land claim, was subject to long-standing dispute with the Northern Territory government. For more on this, see Povinelli, *The Cunning of Recognition*.

110. More recently, LDC has worked with gas company INPEX to provide financial assistance in the form of scholarships, education support, and elder care. Adlam, "Larrakia and INPEX Forge Solid Partnership."

111. Larrakia Nation, "Services."

112. NTASB, "FAQ."

113. Larrakia Nation, "App Gives Culturally Appropriate Support for Tackling Anti-Social Behaviour."

114. Larrakia Nation, "App Gives Culturally Appropriate Support for Tackling Anti-Social Behaviour."

115. NTASB, "Understanding ASB."

116. In the Northern Territory, "disorderly" conduct involving noise remains criminalized under the Summary Offences Act 1923: prohibited behaviors include singing an obscene song or ballad, creating "undue" noise though a social gathering after midnight, making noise after being asked to stop by the police, making noise that would reasonably cause substantial annoyance to others, and playing a musical instrument in public.

117. For more on postrevanchism and the combination of punitive and "compassionate" approaches, see DeVerteuil, "Post-Revanchist Cities?"; Murphy, "'Compassionate' Strategies of Managing Homelessness." Likewise, some scholars have noted a gap between revanchist rhetoric and public policy and its enactment "on the ground." For example, see Johnsen and Fitzpatrick, "Revanchist Sanitisation or Coercive Care?" Within an Australian context, Andrew Clarke and Cameron Parsell contend that, despite experiencing extensive neoliberalization, Australian cities have not embraced revanchist politics to the extent of their North American counterparts, "meaning there has been greater scope for care rationalities to shape homeless governance." See Clarke and Parsell, "The Ambiguities of Homelessness Governance," 1625.

118. Fisher, "Running Amok or Just Sleeping Rough?"

119. Fisher, "Becoming the State in Northern Australia," 239.

120. Fisher, "Becoming the State in Northern Australia."

121. Darwin City Council gave permission for the private security force to operate but only on council-owned land. The operation of the private security forces seemingly exceeds this remit, however. *ABC News*, "NT Security Guards Unlawfully Moving on Sleeping Homeless People in Darwin."

122. Legal groups have argued that the government's employment of private security forces (whose conduct is subject to even less scrutiny than police officers) to forcibly remove outdoor sleepers leaves unaddressed the underlying factors that contribute to homelessness in Darwin, including the lack of access to life-sustaining resources within remote communities, such as housing, education, healthcare, and safe, accessible short- and long-term accommodation in the territory's urban centers. See *ABC News*, "The City Paying Private Security to Patrol the Streets Like Police."

123. Northern Territory Government, "Reduce Crime: New Laws Introduced."

124. Australian Institute of Health and Welfare, "Youth Justice in Australia 2023–24, Unsentenced Detention."

125. Jacquie Wang has discussed the role of automation in policing as a means of extraction and in the context of police department budget cuts. See Wang, *Carceral Capitalism*.

126. Hall et al., *Policing the Crisis*, 390.

127. For a pertinent discussion of the political consciousness and agency of racialized surplus populations see Hall et al., *Policing the Crisis*; Alves, *The Anti-Black City*.

## CHAPTER 5. LULL, MASK, CALM, RENEW

1. Max Richter, "From Sleep."

2. Max Richter, "Sleep."

3. In positioning aesthetic complexity as repellent, and attaching classical music's success to its audience appeal, resonances emerge here between Ritcher's articulation of and proposed remedy for classical music's "modernist" decline and some of the composers discussed in Marianna Ritchey's astute analysis of classical music's incorporation of neoliberal market principles. See Ritchey, *Composing Capital.*

4. Dodero, "Ode to (Proper) Sleep."

5. Crary, *24/7*, 126.

6. Crary, *24/7*, 126–28.

7. Brown, "Review: The All-Night, Outdoor Concert 'Sleep' Creates a Calming Reprieve with a Sense of Loss."

8. Morandeira, "Politics of Sleep: A Text for the Right to Rest," 123.

9. Thompson, "Time, Work-Discipline, and Industrial Capitalism."

10. Morandeira, "Politics of Sleep—A Text for the Right to Rest."

11. Crary, *24/7*; Melbin, *Night as Frontier.*

12. Escobar et al., "Development of the Circadian System and Relevance of Periodic Signals for Neonatal Development."

13. Colesanti et al., *Submerged Literature in Ancient Greek Culture: An Introduction.*

14. Nzengou-Tayo, "Haitian Callaloo."

15. Hillers, "Dialogue or Monologue?"

16. The US ethnomusicologist Natalie Curtis notes in her 1921 account of "American Indian cradle songs" that fathers as well as mothers sing lullabies, amid various remarks about Native American paternal attitudes towards children and involvement in childrearing. See Curtis, "American Indian Cradle Songs."

17. Thompson, "The (Non-)Gendered Practice of Irish Traditional Lullabies."

18. Pester, "Songs of Rest."

19. Hillers, "Dialogue or Monologue?," 37.

20. Hillers, "Dialogue or Monologue?," 53.

21. Cited in Drott and Thompson, "'Is Your Baby Getting Enough Music?,'" 136.

22. Palmer Wolf, *Lullaby: Being Together, Being Well*, 15.

23. Ascenso, Sara, *The Lullaby Project*, 12–13.

24. Brandon et al., "A History of the Theory of Prenatal Attachment," 202.

25. Bowlby, *Maternal Care and Mental Health*, 11.

26. Vicedo, *The Nature and Nurture of Love*, 9–10.

27. Increased demand, recruitment issues, and a lack of funding security have impacted access to and availability of specialist services, with more than 30,000 people who are pregnant or recently postpartum on NHS waiting lists as of August 2023. See Walker, "Sharp Rise in Wait Times for Perinatal Mental Health Care in England." See also chapter 3 for a discussion of music's use as a cheap resource in perinatal health services.

28. See also Drott and Thompson, "'Is Your Baby Getting Enough Music?'"

29. Bernini and Mormando, *The Life of Gian Lorenzo Bernini*, 217.

30. Riva et al., "Gian Lorenzo Bernini's 17th Century White Noise Machine."

31. Riva et al., "Gian Lorenzo Bernini's 17th Century White Noise Machine," 776.

32. Broadbent, "A Clinical Lecture on Sleeplessness," 215; see also Shuttleworth, "Fagged Out."

33. Beaumont, "Insomnia and the Late Nineteenth-Century Insomniac," 2.

34. Beaumont, "Insomnia and the Late Nineteenth-Century Insomniac," 3.

35. Beaumont, "Insomnia and the Late Nineteenth-Century Insomniac," 3.

36. Beaumont, "Insomnia and the Late Nineteenth-Century Insomniac," 4.

37. Shuttleworth, "Fagged Out."

38. A 1558 letter from Lord Talbot to Lord Burghley, which accompanied the delivery of a substance taken from the latter's household, provides details on how to take the substance in order to procure sleep. See Hunter, "'To Cause Sleepe Safe and Shure.'"

39. Shuttleworth, "Fagged Out."

40. Robson-Mainwaring, "Electro-Therapeutic Cures in the Victorian Age."

41. A 1894 patent by Per Erik Fahlberg and titled "Improvements in apparatus for producing a sleep-giving or appeasing sound" describes a device that produces a "softly whispering sound that tends to produce sleep." See Falhberg, Improvements in apparatus for producing a sleep-giving or appeasing sound.

42. Buckwalter, Sleep-inducing sound-producing device; Beazley, Acoustic sleep-inducing apparatus; Wiesman, Device for inducing sleep and rest.

43. Hagood, *Hush*, 109.

44. Hagood, *Hush*, 101–2.

45. Children and infants are similarly absent from Anahid Kassabian's 2013 analysis of sound-based sleep aids, beyond a brief mention of the app Baby Sleep. Kassabian frames these as technologies of self-management, situating them in relation to discourses on the quantified self. However, Kassabian is careful to note that while they my be devices that allow management of the self, this is not to assume management *by* the self. See Kassabian, "Music for Sleeping."

46. For instance, Leap Fitness Group's Sleep Sounds app features a lullaby music box option, while Yogasleep's Nod speaker features multiple instrumental lullaby tracks, mainly played on piano.

47. For more on the orientalism of sleep apps see Kassabian, "Music for Sleeping." While this orientalism is arguably less prevalent now than it was in 2013, when Kassabian was writing, it is still recognizable from many apps. The app Deep Sleep Music: Sleep Sounds by the developer Sleep Sounds App, which allows user to mix sounds, is instructive. A section titled "oriental" features looped sounds of bowls, gongs, bells, guzheng, didgeridoo, and an *om* chant. Among the

options available under the "relaxing" section are musical loops of piano, guitar, saxophone and "Native American flute."

48. Morphée, Morphée Sleep Aid.

49. Morphée, My Little Morphée.

50. For a related analysis of ambient media and its relationship to self-care, see Haffke, "'Adaptive Environments.'"

51. This formulation echoes Anahid Kassabian's own description of what she describes as "ubiquitous listening," which she associates with, among other things, the rise of mobile and digital technologies. See Kassabian, *Ubiquitous Listening*.

52. Kokoon, Philips x Kokoon Sleep Headphones!

53. Raphael, "Calm's Michael Acton Smith Has Come to Hollywood to Put Audiences to Sleep."

54. Raphael, "Calm's Michael Acton Smith Has Come to Hollywood to Put Audiences to Sleep." The "standard" of the three-minute pop song is also traceable to record formats, naturalized with the 78-rpm shellac record but persisted beyond the medium's dominance. See Björnberg, "Why 3 Minutes?"

55. Hissong, *Post Malone, Ariana Grande, Shawn Mendes Will Drop Sleep Remixes.*

56. Hissong, *Post Malone, Ariana Grande, Shawn Mendes Will Drop Sleep Remixes.*

57. Calm, "Calm: The #1 App for Meditation and Sleep."

58. "Calm for Organizations"; Calm Health, "Digital Mental Health App for Health Plans."

59. Haque, "Meditation App Calm Sacks One-Fifth of Staff."

60. Espada, "Calm Is One of the 2022 *Time*100 Most Influential Companies."

61. Acton Smith, *The Magic of Sleep.*

62. Universal Music Group, "Calm and Universal Music Group Create New Sleep Mixes."

63. Hissong, *Post Malone, Ariana Grande, Shawn Mendes Will Drop Sleep Remixes.*

64. soundBrilliance, "soundBrilliance™."

65. Sollos, "Sollos."

66. Sollos, "Technology"; Sollos, "Can Music Make You Feel Better?" There are some notable similarities between Universal's emphasis on "human"-made music, and the ways that other music streaming services have sought to differentiate themselves from competitors. Nick Seaver's discussion of Beats streaming service as offering "artist curation" in opposition to "algorithmic curation" is instructive in this regard. Seaver, *Computing Taste.*

67. Universal Music Group, "Universal Music Group and Thrive Global Launch Music + Health Summit."

68. Universal has been partnering with various AI companies, while Sollos's framing of AI as something that both threatens musical quality and can

be effectively incorporated into music production speaks to Universal's ongoing attempts to position the company as arbiters of "quality" in the context of "AI slop." For example, Universal described their partnership with NVIDIA as enabling "responsible AI for music creation, delivery and engagement." Such framing speaks to the company's desire to exploit the economic opportunities offered with the rise in AI investment while also protecting their existing revenue streams. See Stassen, "UMG's Latest Major AI Partnership Arrives via Tech Giant NVIDIA, with Promise of 'Antidote to Generic AI Slop.'"

69. Rappaport, "The Global Wellness Industry Is Now Worth $6.3 Trillion."

70. Salmenniemi, "The Crisis of Social Reproduction and Therapeutic Self-Care."

71. Drott, *Streaming Music, Streaming Capital*, 1–2.

72. Negus, "From Creator to Data."

73. Olson, "Meeting of The Minds: Why The Music & Mental Wellness Industries Are Syncing Up Now More Than Ever."

74. Laing, "A Voice without a Face," 6–7.

75. Hesmondhalgh and Meier, "What the Digitalisation of Music Tells Us about Capitalism, Culture and the Power of the Information Technology Sector."

76. Hesmondhalgh and Meier, "What the Digitalisation of Music Tells Us about Capitalism, Culture and the Power of the Information Technology Sector."

77. Negus, "From Creator to Data."

78. Mozilla Foundation, "Privacy Not Included: Calm."

79. Mozilla Foundation, "Privacy Not Included: Calm."

80. Amplitude, "Breathe Deeply and Disrupt: How Calm Uses Analytics to Improve Sleep, Health and Happiness."

81. Prey, "Musica Analytica," 32.

82. Hissong, "Calm's Head of Music Courtney Phillips—Future 25."

83. Acton Smith, *The Magic of Sleep*.

84. Amplitude, "Breathe Deeply and Disrupt: How Calm Uses Analytics to Improve Sleep, Health and Happiness."

85. The designation of "service" here differs from the definition provided by Benanav and Clegg, discussed in the book's introduction. Unlike services provided by human labor, which, as Benanav and Clegg suggest, remain formally subsumed, digital services are more akin to goods in terms of their production. However, for the purpose of this chapter, they are distinguished from goods in the sense they are rented rather than owned. This will be discussed further below.

86. As the Endnotes collective make clear, socially reproductive activities can be both directly and indirectly market mediated, and they may or may not be subject to rationalization, depending on their socioeconomic context. Endnotes, "The Logic of Gender."

87. For example, see Drott, *Streaming Music, Streaming Capital*; Rigi and Prey, "Value, Rent, and the Political Economy of Social Media." For a critique of

this analytic move relating to digitized musical reproduction and its relationship to value see Parkhurst, "Digital Information and Value."

88. Fleischer, "If the Song Has No Price, Is It Still a Commodity?"

89. Johnson, "App Store Tip: Beware of Knockoffs."

90. Srnicek, *Platform Capitalism.*

91. Smith, "Rethinking Sleep."

92. Nudelman, *Fighting Sleep.*

93. Blank Noise, "About Meet to Sleep."

94. The Nap Ministry, "One Woman Shows: Transfiguration and Reparations LIVE!"

95. Good, "A Playlist for Rest, Curated by The Nap Ministry's Tricia Hersey."

96. Hersey has rejected the term "self-care" and instead uses the phrase "soul care," insofar as the former has become depoliticized through its extraction from its Black feminist origins. For on the depoliticization of self-care and Hersey's definition of soul care see Monier, "Rest as Resistance."

## EPILOGUE

1. Valle, "Trad Values Meets Tech."

2. Sherman, "The Rise of Pronatalism." Pronatalism has also held influence outside the US right. For example, it has informed the policies of the populist right Orbán administration in Hungary, with state benefits available to (specifically) heterosexual, married couples earning above minimum wage who are raising children.

3. Behrensen, "Tech Oligarchs and the Rise of Silicon Valley Pronatalism."

4. For more on the proximity of white supremacy and ethnonationalism to US pronatalism, see Munn, "Pronatalism Is the Latest Silicon Valley Trend: What Is It—and Why Is It Disturbing?'; Behrensen, "Tech Oligarchs and the Rise of Silicon Valley Pronatalism"; Sherman, "The Rise of Pronatalism."

5. Cammack, "The Political Economy of Post-Reproduction Society."

6. Cammack, "The Political Economy of Post-Reproduction Society."

7. Weinbaum, *The Afterlife of Reproductive Slavery*; Davis, *Women, Race, & Class*; Nash, *Birthing Black Mothers*; Lewis, *Full Surrogacy Now.*

8. Nash, *Birthing Black Mothers*, 3.

9. Reynolds, "Abortion Is Killing the Black Community."

10. Kelly, "Fact Check."

11. Whitmarsh, "Disinformation and the Ghost of Margaret Sanger." For other examples of the mobilization of racism in anti-abortion attacks on Planned Parenthood, see Browne, "Anti-Abortion Feminism."

12. The myth of abortion as a means of exterminating the Black population has been readily and thoroughly critiqued by Black feminists. For example, see Dobbins-Harris, "The Myth of Abortion as Black Genocide."

13. Dobbins-Harris, "The Myth of Abortion as Black Genocide." For more on the formation on reproductive justice as a concept and its priorities, see Ross et al., *Radical Reproductive Justice*.

14. For an example of reporting on Carson's report, see Fessler, "Housing Secretary Ben Carson Says Poverty Is a 'State Of Mind.'"

15. Endnotes, "Gender, Race, Class and Other Misfortunes."

16. Brinkhurst-Cuff, "Nadia Rose's 'Skwod' Used to Ward Off Fascists in London."

17. Of significance here, too, is the attack on Black kinship relations under slavery, which excluded enslaved people from the (white) institution of the family: for more on this see Davis, *Women, Race, & Class*; O'Brien, *Family Abolition*. The notion of Black women in popular music offering—and indeed enacting—a life outside of capitalist reproductive norms also finds resonance in Daphne A. Brook's account of Black feminist sound, in which she highlights the "fugitive" lives forged and expressed by Black women musicians. See Brooks, *Liner Notes for the Revolution*.

18. For more on the nostalgic revisitation of the dancefloor and its omission of friction, see Goh and Thompson, *Sonic Cyberfeminisms* [forthcoming].

19. For analyses of the politics of the ballroom scene see Roberson, "Ballroom: The Trans Sounds of Black Freedom"; Solanes Valldaura, "MOTHER HAS ARRIVED! Queer Kinship, Care Practices and Reconfigurations of (M)Otherhood within Ballroom Culture."

20. Gilroy, *There Ain't No Black in the Union Jack*, 247.

21. Gilroy's framing of the "night time as the right time" also speaks to recent anxieties about the nighttime economy's decline. Dan Hancox offers an important critique of tropes about the "death" of London's nightlife, as well as some of the research that has been produced to support such claims. At the same time, Hancox rightfully notes the detrimental implications of the nighttime economy for workers. See Hancox, "Debunking the Weird Myths about London's 24-Hour Party People."

22. For example, education and care programs were run out of LA's Black lesbian and queer nightclub Catch One, founded by Jewel Thais-Williams. Brown, "Jewel Thais-Williams, Founder of Beloved Black Queer Nightclub Jewel's Catch One, Dies at 86." Songs, choirs, and musical performance have played an important role in AIDS activism in South Africa and elsewhere: for more on this see Okigbo, "Performing Blackness in the South African HIV/AIDS Choir"; Vokwana, "Sinikithemba Gospel Group and the Grassroots Struggle against HIV/AIDS Stigma in South Africa"; Walker, *Songs of the South African AIDS Crisis*.

23. Walker, "Washington DC's Go-Go Music Hits Back in Fight against Gentrification"; Demony et al., "Madrid Residents Fight against Evictions with Musical Protest."

24. For an astute analysis of the segregation of (popular) music from wage-time, see Rekret, *Take This Hammer: Work, Song, Crisis*.

# Bibliography

ABC News. "NT Security Guards Unlawfully Moving on Sleeping Homeless People in Darwin." 5 November 2023. https://www.abc.net.au/news/2023-11-06/nt-private-security-guards-unlawfully-moving-on-homeless-people/103060992.

ABC News. "The City Paying Private Security to Patrol the Streets Like Police." 29 October 2023. https://www.abc.net.au/news/2023-10-30/private-security-policing-darwin-city-four-corners/103013202.

Acton Smith, Michael. *The Magic of Sleep: Talks at Google*. 2020. https://www.youtube.com/watch?v=6dq9qIuY4GA.

Adkins, Lisa, Martijn Konings, and Melinda Cooper. *The Asset Economy: Property Ownership and the New Logic of Inequality*. Polity Press, 2020.

Adlam, Nigel. "Larrakia and INPEX Forge Solid Partnership." *Territory Q*, 1 October 2020. https://territoryq.com.au/larrakia-and-inpex-forge-solid-partnership/.

Adolph, Carolyn, and Joshua McNicols. "Meet the Coders Who Hear You Yell at Alexa." KUOW, 25 March 2019. https://www.kuow.org/stories/meet-the-foot-soldiers-of-the-wake-word-war.

Agar, Nicholas. "Liberal Eugenics." *Public Affairs Quarterly* 12, no. 2 (1998): 137–55.

Alexa Skills Kit. "What Is Conversational AI?" Accessed 14 March 2025. https://developer.amazon.com/en-GB/alexa/alexa-skills-kit/conversational-ai.

Allen Anderson, Paul. "Neo-Muzak and the Business of Mood." *Critical Inquiry* 41, no. 4 (2015): 811–40. https://doi.org/10.1086/681787.

Alli, Sabrina. "Carceral Educations." *The New Inquiry*, 22 September 2014. https://thenewinquiry.com/carceral-educations/.

Alves, Jaime Amparo. "Our Fire This Time! The Favela As Zone Of Death And Insurgent Spatiality." *The Funambulist*, 9 March 2018. https://thefunambulist.net/magazine/16-proletarian-fortresses/fire-time-favela-zone-death-insurgent-spatiality-jaime-alves.

Alves, Jaime Amparo. *The Anti-Black City: Police Terror and Black Urban Life in Brazil*. University of Minnesota Press, 2018.

Alzheimer's Disease International. *World Alzheimer Report 2019: Attitudes to Dementia*. Alzheimer's Disease International, 2019.

Amazon. "Divorced Siri: Customer Review." Accessed 14 March 2025. https://www.amazon.com/gp/customer-reviews/R5C6UBI2YCFEQ/ref=cm_cr_srp_d_rvw_ttl?ie=UTF8&ASIN=B01DFKC2SO.

Amazon. "Help: Interest-Based Ads." Accessed 3 September 2020. https://www.amazon.co.uk/gp/help/customer/display.html?nodeId=201909150.

Amazon. "Help: Privacy Notice." Accessed 3 September 2020. https://www.amazon.co.uk/gp/help/customer/display.html?nodeId=201909010#GUID-A440AA65-7F7E-4134-8FA8-842156F43EEE__SECTION_22160257376047E78334D565CD73852D.

Amazon Advertising. "Amazon DSP (Demand Side Platform)." Accessed 21 September 2020. https://advertising.amazon.com/solutions/products/amazon-dsp?ref_=a20m_us_libr_bos_intro_dsp.

Amazon Alexa "Sharing is caring." 2019. https://www.youtube.com/watch?v=225Wlg3pkdo.

Amazon News. "Alexa Can Now Help Make the Remote Caregiving Experience Easier." 11 November 2010. https://www.aboutamazon.com/news/devices/alexa-can-now-help-make-the-remote-caregiving-experience-easier.

Amazon Web Services. "Alexa for Business." Accessed 14 March 2023. https://aws.amazon.com/alexaforbusiness/.

Amplitude. "Breathe Deeply and Disrupt: How Calm Uses Analytics to Improve Sleep, Health and Happiness." n.d. https://info.amplitude.com/FY21Q2-Event-Digital-Disruptors-Summit_01b-Confirmation-video-layout?wvideo=ngqr1vqh7b.

Anderson, Bridget. *Doing the Dirty Work?: The Global Politics of Domestic Labour*. Palgrave Macmillan, 2000.

Andrejevic, Mark. *Infoglut: How Too Much Information Is Changing the Way We Think and Know*. Routledge, 2013.

Andrews, Travis M. "Alexa, Just Shut Up." *Washington Post*, 1 July 2020. https://www.washingtonpost.com/technology/2020/07/01/alexa-siri-google-home-assistant/.

Anti-Social Behaviour Act 2003 (2003). https://www.legislation.gov.uk/ukpga/2003/38/section/29.
Arcy, Jacquelyn. "Emotion Work: Considering Gender in Digital Labor." *Feminist Media Studies* 16, no. 2 (2016): 365–68. https://doi.org/10.1080/14680777.2016.1138609.
Armstrong, Andrea. "Creating Sustainable Communities in 'Newcastle-Gateshead.'" PhD dissertation Durham University, 2010. http://etheses.dur.ac.uk/393/.
Aronson, Jane, and Sheila M. Neysmith. "'You're Not Just in There to Do the Work': Depersonalizing Policies and the Exploitation of Home Care Workers' Labor." *Gender and Society* 10, no. 1 (1996): 59–77.
Ascenso, Sara. *The Lullaby Project: Areas of Change, Mechanisms of Impact.* The Irene Taylor Trust, 2017.
Atanasoski, Neda, and Kalindi Vora *Surrogate Humanity: Race, Robots, and the Politics of Technological Futures.* Duke University Press, 2019.
Australian Institute of Health and Welfare. "Youth Justice in Australia 2023–24, Unsentenced Detention." 28 March 2025. https://www.aihw.gov.au/reports/youth-justice/youth-justice-in-australia-2023-24/contents/detention/unsentenced-detention.
Baby Doppler. "Sonoline B Fetal Doppler." Accessed 12 March 2025. https://www.babydoppler.com/gb/en-gb/sonoline-b-baby-doppler-fetal-monitor-teal.html.
BabyPlus. "About." Accessed 12 March 2025. https://babyplus.com/about/.
BabyPlus. "BabyPlus® Prenatal Education System." Accessed 12 March 2025. https://babyplus.com/product/baby-plus/.
BabyPlus. "FAQs." Accessed 12 March 2025. https://babyplus.com/faqs/#1590112333732-3da04ce6-3fce.
BabyPlus. "Supporting Science." Accessed 12 March 2025. https://babyplus.com/supporting-science/.
The BabyPlus Company USA. *Prenatal Education: When Learning Begins with BabyPlus.* 4 June 2013. https://www.youtube.com/watch?v=SouFlnrgakQ.
The BabyPlus Company USA, *Prenatal Education: When Learning Begins with BabyPlus.* 4 June 2013. https://www.youtube.com/watch?v=SouFlnrgakQ.
Babypod. "What Is Babypod?" Archived 26 August 2019, at http://web.archive.org/web/20190826001102/https://babypod.net/en/what-is-babypod/.
Babypod. "What Music Do Foetuses Prefer?" Accessed 12 March 2025. https://babypod.net/en/blog/what-music-do-foetuses-prefer/.
Backer, David I., and Kate Cairns. "Social Reproduction Theory Revisited." *British Journal of Sociology of Education* 42, no. 7 (2021): 1086–104. https://doi.org/10.1080/01425692.2021.1953962.
Bailey, Peter. *Leisure and Class in Victorian England: Rational Recreation and the Contest for Control, 1830–1885.* Routledge, 2014.

Baker, Geoffrey. *El Sistema: Orchestrating Venezuela's Youth*. Oxford University Press, 2014.

Barbara, Vanessa. "Whose Mall Is It?'" *The New York Times*, 16 January 2014. https://www.nytimes.com/2014/01/16/opinion/barbara-whose-mall-is-it.html?_r=1.

Battistoni, Alyssa. "Ideology at Work? Rethinking Reproduction." *American Political Science Review* 119, no. 3 (2025): 1205–18. https://doi.org/10.1017/S0003055424000868.

Baynton, Douglas C. *Forbidden Signs: American Culture and the Campaign against Sign Language*. University of Chicago Press, 1996.

*BBC News*. "Symphonies Soothe Train Vandals." BBC News, 30 January 1998. http://news.bbc.co.uk/1/hi/uk/51939.stm.

*BBC News*. "Why Is a Council Giving Out Virtual Assistants?" n.d. Accessed 14 March 2025. https://www.bbc.co.uk/news/av/uk-politics-43869120.

*BBC News*. "Wrexham McDonald's Plays Classical Music to Deter Bad Behaviour." *BBC News*, 30 January 2023. https://www.bbc.co.uk/news/uk-wales-64458332.

Beaumont, Matthew. "Insomnia and the Late Nineteenth-Century Insomniac: The Case of Albert Kimball." *Interface Focus* 10, no. 3 (2020): 20190074. https://doi.org/10.1098/rsfs.2019.0074.

Beauvais, Clémentine. "The 'Mozart Effect': A Sociological Reappraisal." *Cultural Sociology* 9, no. 2 (2015): 185–202. https://doi.org/10.1177/1749975514557096.

Beazley, Charles. Acoustic sleep-inducing apparatus. Patent US2644153A, filed 12 April 1950, and issued 30 June 1953. https://patents.google.com/patent/US2644153A/en.

Becker, Gary S., ed. *A Treatise on the Family*. Harvard University Press, 2010.

Beech, Dave. *Art and Postcapitalism: Aesthetic Labour, Automation and Value Production*. Pluto Press, 2019.

Behrensen, Maren. "Tech Oligarchs and the Rise of Silicon Valley Pronatalism." *Tech Policy*. Accessed 25 September 2025. https://www.techpolicy.press/tech-oligarchs-and-the-rise-of-silicon-valley-pronatalism/.

Benanav, Aaron. *Automation and the Future of Work*. Verso, 2020.

Benanav, Aaron, and John Clegg. "Misery and Debt: On the Logic and History of Surplus Populations and Surplus Capital." In *Contemporary Marxist Theory: An Anthology*, edited by Andrew Pendakis, Jeff Diamanti, Nicholas Brown, Josh Robinson, and Irme Szeman. Bloomsbury, 2014.

Bernini, Domenico, and Franco Mormando. *The Life of Gian Lorenzo Bernini*. Penn State Press, 2012.

"Best Super Affordable Alexa Devices for 2020." CNET. Accessed 2 September 2020. https://www.cnet.com/news/best-super-affordable-alexa-devices-for-2020/.

Bhattacharya, Snehashish, Surbhi Kesar, and Sahil Mehra. "Exclusion, Surplus Population, and the Labour Question in Postcolonial Capitalism: Future Directions in Political Economy of Development." *Review of Political Economy* 35, no. 1 (2023): 145–73. https://doi.org/10.1080/09538259.2022.2134650.

Bhattacharya, Tithi, ed. *Social Reproduction Theory: Remapping Class, Recentering Oppression*. Pluto Press, 2017.

Björnberg, Alf. "Why 3 Minutes? Sound-Reproduction Media, Popular Music Form and Temporality." *Svensk Tidskrift För Musikforskning / Swedish Journal of Music Research* 102 (December 2020): 1–14.

Blank Noise. "About Meet To Sleep." 2023. https://www.blanknoise.org/aboutmeettosleep.

Bohn, Dieter. "Amazon Says 100 Million Alexa Devices Have Been Sold." *The Verge*, 4 January 2019. https://www.theverge.com/2019/1/4/18168565/amazon-alexa-devices-how-many-sold-number-100-million-dave-limp.

Bonnisseau, Nathan. "Ostentação: A Game of Power for Politicians and Funk Artists." RioOnWatch, 29 October 2014. https://www.rioonwatch.org/?p=18142.

Booth, Jessica. "12 Self-Care Playlist Ideas to Listen to When You Just Need Time for Yourself." Bustle, 9 August 2018. https://www.bustle.com/p/12-self-care-playlist-ideas-to-listen-to-when-you-just-need-time-for-yourself-10054502.

Bourgault, Sophie. "Attentive Listening and Care in a Neoliberal Era: Weilian Insights for Hurried Times." *Politics and Ethics* 18, no. 3 (2016): 311–77.

Bowers, Nathan David. "Creating a Home Culture for the Phonograph: Women and the Rise of Sound Recordings in the United States, 1877–1913." PhD, University of Pittsburgh, 2007.

Bowlby, John. *Maternal Care and Mental Health: A Report Prepared on Behalf of the World Health Organization as a Contribution to the United Nations Programme for the Welfare of Homeless Children*. 2nd edition. World Health Organization, 1952.

Brandon, Anna R., Sandra Pitts, Wayne H. Denton, C. Allen Stringer, and H. M. Evans. "A History of the Theory of Prenatal Attachment." *Journal of Prenatal & Perinatal Psychology & Health : APPPAH* 23, no. 4 (2009): 201–22.

Brinkhurst-Cuff, Charlie. "Nadia Rose's 'Skwod' Used to Ward Off Fascists in London." *Dazed*, 7 May 2017. https://www.dazeddigital.com/music/article/35827/1/nadia-roses-skwod-used-to-ward-off-fascists-in-croydon.

Broadbent, William H. "A Clinical Lecture on Sleeplessness." *The Lancet* 155, no. 3987 (1900): 215–17. https://doi.org/10.1016/S0140-6736(01)70176-6.

Brooks, Daphne A. *Liner Notes for the Revolution: The Intellectual Life of Black Feminist Sound*. Belknap Press of Harvard University Press, 2021.

Brown, August. "Jewel Thais-Williams, Founder of Beloved Black Queer Nightclub Jewel's Catch One, Dies at 86." *Los Angeles Times*, 9 July 2025. https://www.latimes.com/entertainment-arts/music/story/2025-07-08/jewel-thais-williams-obituary-jewels-catch-one-los-angeles-dies-86.

Brown, August. "Review: The All-Night, Outdoor Concert 'Sleep' Creates a Calming Reprieve with a Sense of Loss." 29 July 2018. https://www.latimes.com/entertainment/music/la-et-ms-0730-sleep-live-20180727-story.html.

Brown, Wendy. *Undoing the Demos: Neoliberalism's Stealth Revolution.* Zone Books, 2017.

Browne, Victoria. "Anti-Abortion Feminism: How Is This Even a Thing?" *Radical Philosophy*, no. 213 (2022): 27–42.

Buck-Morss, Susan. "The Flaneur, the Sandwichman and the Whore: The Politics of Loitering." *New German Critique*, no. 39 (1986): 99–140. https://doi.org/10.2307/488122.

Buckwalter, James K. Sleep-inducing sound-producing device. Patent US3121220A, filed 21 January 1963, and issued 11 February 1964. https://patents.google.com/patent/US3121220A/en.

Bueno, Claudio Celis. *The Attention Economy: Labour, Time and Power in Cognitive Capitalism.* Rowman & Littlefield, 2016.

Bull, Anna. *Class, Control, and Classical Music.* Oxford University Press, 2019.

Caldeira, Teresa Pires do Rio. *City of Walls: Crime, Segregation, and Citizenship in São Paulo.* University of California Press, 2020.

Calm. "Calm: The #1 App for Meditation and Sleep." Accessed 12 January 2024. https://www.calm.com/.

"Calm for Organizations." *Calm Health.* Accessed 25 January 2026. https://health.calm.com/calm-for-organizations/.

Calm Health. "Digital Mental Health App for Health Plans." Accessed 8 January 2026. https://health.calm.com/health-plans/.

Cammack, Paul. "The Political Economy of Post-Reproduction Society." *New Political Economy* (2025): 1–16. https://doi.org/10.1080/13563467.2025.2555352.

Cammack, Paul, and Martha E. Gimenez. "The Permanent Global Crisis of Working-Class Social Reproduction: Ten Propositions." *Global Political Economy* 4, no. 2 (2025): 104–19. https://doi.org/10.1332/26352257Y2024D000000025.

Capgemini. "Smart Talk: How Organizations and Consumers Are Embracing Voice and Chat Assistants." Capgemini Research Institute, 2019. https://www.capgemini.com/news/press-releases/smart-talk-how-organizations-and-consumers-are-embracing-voice-and-chat-assistants/.

Cardoso, Leonardo. *Sound-Politics in São Paulo.* Oxford University Press, 2019.

Chan, Jenny, Mark Selden, and Pun Ngai. *Dying for an iPhone: Apple, Foxconn and the Lives of China's Workers.* Pluto Press, 2020.

Chan, Rosalie. "Amazon Web Services Generated over $10 Billion in Quarterly Revenue." *Business Insider*, 1 April 2020. https://www.businessinsider.com/amazon-earnings-aws-amazon-web-services-10-billion-quarterly-revenue-2020-4?r=US&IR=T.

Christophers, Brett. *Rentier Capitalism: Who Owns the Economy, and Who Pays for It?* Verso, 2020.

Clarke, Andrew, and Cameron Parsell. "The Ambiguities of Homelessness Governance: Disentangling Care and Revanchism in the Neoliberalising City." *Antipode* 52, no. 6 (2020): 1624–46. https://doi.org/10.1111/anti.12671.

Clifford Rosenberg, Buck "Shhh! Noisy Cities, Anti-Noise Groups and Neoliberal Citizenship." *Journal of Sociology* 52, no. 2 (2016): 190–203. https://doi.org/10.1177/1440783313507493.

Colesanti, Giulio, Manuela Giordano, and Maria Broggiato. *Submerged Literature in Ancient Greek Culture: An Introduction*. Submerged Literature in Ancient Greek Culture, volume 1. De Gruyter, 2014. https://doi.org/10.1515/9783110334081.

Community Care. "Talking Alexa: How Assistive Technology Can Help Adult Social Care Services." 18 October 2018. https://www.communitycare.co.uk/2018/10/18/talking-alexa-assistive-technology-can-help-adult-social-care-services/.

Cooper, Melinda. *Family Values: Between Neoliberalism and the New Social Conservatism*. Zone Books, 2017.

Cowan, Alexander. "Unsound: A Cultural History of Music and Eugenics." Harvard University, 2023. https://dash.harvard.edu/entities/publication/dacd399f-53fe-496e-820d-aaf209ba77d5.

Cowan, Ruth Schwartz. *More Work for Mother: The Ironies of Household Technology from the Open Hearth to the Microwave*. Basic Books, 1983.

Cox, Nicole, and Silvia Federici. *Counter-Planning from the Kitchen*. Falling Wall Press, 1975.

Crary, Jonathan. *24/7: Late Capitalism and the Ends of Sleep*. Verso, 2013.

Crawford, Kate. *Atlas of AI: Power, Politics, and the Planetary Costs of Artificial Intelligence*. Yale University Press, 2021.

Crawford, Kate, and Vladan Joler. *Anatomy of an AI System: Amazon Echo as an Anatomical Map of Human Labour, Data and Planetary Resources*. Share Lab and AI Now, 2018. www.anatomyof.ai.

Curtis, Natalie. "American Indian Cradle Songs." *The Musical Quarterly* 7, no. 4 (1921): 549–58. https://doi.org/10.1093/mq/VII.4.549.

Daar, Judith. *The New Eugenics: Selective Breeding in an Era of Reproductive Technologies*. Yale University Press, 2017.

Dalla Costa, Maria, and Selma James. "Women and the Subversion of the Community." In *Materialist Feminism: A Reader in Class, Difference, and Women's Lives*, edited by Rosemary Hennessy and Chrys Ingraham. Routledge, 2008.

Daniels, Cynthia R., and Erin Heidt-Forsythe. "Gendered Eugenics and the Problematic of Free Market Reproductive Technologies: Sperm and Egg Donation in the United States." *Signs: Journal of Women in Culture and Society* 37, no. 3 (2012): 719–47. https://doi.org/10.1086/662964.

Davis, Angela Y. *Women, Race, & Class*. 1st ed. Random House, 1981.

Day, Matt, Giles Turner, and Natalia Drozdiak. "Thousands of Amazon Workers Listen to Alexa Conversations." *Time*, 11 April 2019. https://time.com/5568815/amazon-workers-listen-to-alexa/.

Demony, Catarina, David Latona, Catarina Demony, and David Latona. "Madrid Residents Fight against Evictions with Musical Protest." *Reuters*, 5 February 2024. https://www.reuters.com/world/europe/madrid-residents-fight-against-gentrification-with-musical-protest-2024-02-05/.

Denning, Michael. *Noise Uprising: The Audiopolitics of a World Musical Revolution*. Verso, 2015.

DeNora, Tia. *Music in Everyday Life*. Cambridge University Press, 2000.

DeVerteuil, Geoff. "Post-Revanchist Cities?" *Urban Geography* 40, no. 7 (2019): 1055–61. https://doi.org/10.1080/02723638.2019.1640034.

Devine, Kyle. *Decomposed: The Political Ecology of Music*. MIT Press, 2019.

Dobbins-Harris, Shyrissa. "The Myth of Abortion as Black Genocide: Reclaiming Our Reproductive Cycle." *National Black Law Journal* 26, no. 1 (2017). https://escholarship.org/uc/item/0988p9xp.

Dodero, Camille. "Ode to (Proper) Sleep." Billboard, P-MRC, 2018. https://www.billboard.com/photos/max-richter-sleep-music-tips-interview/1-ode-to-proper-sleep/.

Dredge, Stuart. "'Everybody's Talkin': Smart Speakers and Their Impact on Music Consumption." British Phonograph Industry and Entertainment Retaillers Association, 2018. https://www.bpi.co.uk/media/2856/bpi-era-insight-session-everybodys-talkin-smart-speakers-and-their-impact-on-music-consumption-mar-2018.pdf.

Dreyfuss, Emily. "The Terrible Joy of Yelling at an Amazon Echo." *Wired*, n.d. Accessed 27 August 2020. https://www.wired.com/story/amazon-echo-alexa-yelling/.

Drott, Eric. *Streaming Music, Streaming Capital*. Duke University Press, 2024.

Drott, Eric, and Marie Thompson. "'Is Your Baby Getting Enough Music?': Musical Interventions into Gestational Labor." *Women and Music: A Journal of Gender and Culture* 26, no. 1 (2022): 125–47. https://doi.org/10.1353/wam.2022.0006.

Duchen, Jessica. "Mind the Bach: Classical Music on the Underground." *The Independent*, 26 March 2008. https://www.independent.co.uk/arts-entertainment/music/features/mind-the-bach-classical-music-on-the-underground-800483.html.

Duffy, Brooke Erin. "The Romance of Work: Gender and Aspirational Labour in the Digital Culture Industries." *International Journal of Cultural Studies* 19, no. 4 (2016): 441–57. https://doi.org/10.1177/1367877915572186.

Dyer-Witheford, Nick. *Cyber-Proletariat: Global Labour in the Digital Vortex*. Digital Barricades: Interventions in Digital Culture and Politics. Between the Lines; Pluto Press, 2015.

Edison Research. "UK Smart Speaker Ownership Outpaces U.S." 21 May 2025. https://www.edisonresearch.com/uk-smart-speaker-ownership-outpaces-u-s/.

Edison, Thomas A. "The Phonograph and Its Future." *The North American Review* 126, no. 262 (1878): 527–36. JSTOR.

Edison, Thomas A. "'The Woman of the Future,' *Good Housekeeping (1912)*." Rutgers University Press, 2020. https://doi.org/10.36019/97808135449 46-065.
Ehrlich, Cyril. *The Piano: A History*. Dent, 1976.
Endnotes. "Gender, Race, Class and Other Misfortunes." *Endnotes* 3 (2013). https://endnotes.org.uk/translations/endnotes-the-logic-of-gender.
Endnotes. "The Logic of Gender." *Gender, Race, Class and Other Misfortunes* 3 (2013). https://endnotes.org.uk/translations/endnotes-the-logic-of-gender.
English Heritage. *Heritage Crime Prevention: A Guide for Owners, Tenants and Managers of Heritage Assets*. English Heritage, 2013.
Escobar, Carolina, Adelina Rojas-Granados, and Manuel Angeles-Castellanos. "Development of the Circadian System and Relevance of Periodic Signals for Neonatal Development." In *Handbook of Clinical Neurology*, vol. 179. Elsevier, 2021. https://doi.org/10.1016/B978-0-12-819975-6.00015-7.
Espada, Mariah. "Calm Is One of the 2022 *Time*100 Most Influential Companies." 30 March 2022. https://time.com/collection/time100-companies-2022/6159416/calm/.
Evens, Aden. *Sound Ideas: Music, Machines, and Experience*. University of Minnesota Press, 2005.
Falhberg, Per Erik. Improvements in apparatus for producing a sleep-giving or appeasing sound. Patent 23002, filed 1894. Accessed 8 January 2026. https://wellcomecollection.org/works/sh2qjuyj/items.
Fancourt, Daisy, and Rosie Perkins. "Could Listening to Music during Pregnancy Be Protective against Postnatal Depression and Poor Wellbeing Post Birth? Longitudinal Associations from a Preliminary Prospective Cohort Study." *BMJ Open* 8, no. 7 (2018): e021251. https://doi.org/10.1136/bmjopen-2017-021251.
Farinati, Lucia, and Claudia Firth. *The Force of Listening*. Doormats 6. Errant Bodies Press, 2017.
Federici, Silvia. "Social Reproduction Theory History, Issues and Present Challenges." *Radical Philosophy* 204 (2019): 55–7.
Fessler, Pam. "Housing Secretary Ben Carson Says Poverty Is a 'State Of Mind.'" National. *NPR*, 25 May 2017. https://www.npr.org/2017/05/25/530068988/ben-carson-says-poverty-is-a-state-of-mind.
Fischer, Claude S. "Gender and the Residential Telephone, 1890–1940: Technologies of Sociability." *Sociological Forum* 3, no. 2 (1988): 211–33. https://doi.org/10.1007/BF01115291.
Fisher, Daniel. "Becoming the State in Northern Australia: Urbanisation, Intra-Indigenous Relatedness, and the State Effect." *Oceania* 83, no. 3 (2013): 238–58. https://doi.org/10.1002/ocea.5023.
Fisher, Daniel. "Running Amok or Just Sleeping Rough? Long-Grass Camping and the Politics of Care in Northern Australia." *American Ethnologist* 39, no. 1 (2012): 171–86. https://doi.org/10.1111/j.1548-1425.2011.01356.x.

Fleischer, Rasmus. "If the Song Has No Price, Is It Still a Commodity?: Rethinking the Commodification of Digital Music." *Culture Unbound* 9, no. 2 (2017): 146–62. https://doi.org/10.3384/cu.2000.1525.1792146.

Fortunati, Leopoldina. *The Arcane of Reproduction: Housework, Prostitution, Labor and Capital.* Autonomedia, 1995.

Foucault, Michel. *The Birth of Biopolitics: Lectures at the Collège de France, 1978–79.* Palgrave Macmillan, 2010.

Fraser, Nancy. "Contradictions of Capital and Care." *New Left Review*, 100 (2016): 99–117.

Friedman, Milton, and Rose D. Friedman. *Capitalism and Freedom: Fortieth Anniversary Edition.* University of Chicago Press, 2002.

Fuchs, Christian. "The Digital Labour Theory of Value and Karl Marx in the Age of Facebook, YouTube, Twitter, and Weibo." In *Reconsidering Value and Labour in the Digital Age*, edited by Eran Fisher and Christian Fuchs. Palgrave Macmillan, 2015. https://doi.org/10.1057/9781137478573_2.

Galton, Francis. *Inquiries into Human Faculty and Its Development.* MacMillan, 1883.

Gambino, Lauren. "Anti-Abortion Activists Rally across US as Third Video Targets Planned Parenthood." US News. *The Guardian*, 28 July 2015. https://www.theguardian.com/us-news/2015/jul/28/planned-parenthood-video-anti-abortion-rally-rand-paul.

Garmany, Jeff, and Matthew A. Richmond. "Hygienisation, Gentrification, and Urban Displacement in Brazil." *Antipode* 52, no. 1 (2020): 124–44. https://doi.org/10.1111/anti.12584.

Garrett, P. M. "Making 'Anti-Social Behaviour': A Fragment on the Evolution of 'ASBO Politics' in Britain." *British Journal of Social Work* 37, no. 5 (2006): 839–56. https://doi.org/10.1093/bjsw/bcl033.

Gilbert, Jeremy. "The Second Wave: The Specificity of New Labour Neo-Liberalism." *Soundings* 26 (2004): 25–45.

Gilmore Wilson, Ruth, and Craig Gilmore. "Beyond Bratton." In *Policing the Planet: Why the Policing Crisis Led to Black Lives Matter*, edited by Jordan T. Camp and Christina Heatherton. Verso, 2016.

Gilroy, Paul. *There Ain't No Black in the Union Jack: The Cultural Politics of Race and Nation.* Hutchinson, 1987.

Gioia, Theodore. "Bach at the Burger King." *Los Angeles Review of Books.* 17 May 2018. https://lareviewofbooks.org/article/bach-at-the-burger-king/.

Gitelman, Lisa. "How Users Define New Media: A History of the Amusement Phonograph." In *Rethinking Media Change*, edited by David Thorburn and Henry Jenkins. MIT Press, 2003. https://doi.org/10.7551/mitpress/5930.003.0007.

Gitelman, Lisa. "Reading Music, Reading Records, Reading Race: Musical Copyright and the U.S. Copyright Act of 1909." *The Musical Quarterly* 81, no. 2 (1997): 265–90.

Gitelman, Lisa. *Scripts, Grooves, and Writing Machines: Representing Technology in the Edison Era*. Stanford University Press, 1999.

Gitelman, Lisa. "Souvenir Foils." In *New Media, 1740–1915*, edited by Lisa Gitelman and Geoffrey B. Pingree. MIT Press, 2003. https://doi.org/10.7551/mitpress/5001.003.0011.

Goh, Annie. "Sounding Situated Knowledges: Echo in Archaeoacoustics." *Parallax* 23, no. 3 (2017): 283–304. https://doi.org/10.1080/13534645.2017.1339968.

Goh, Annie and Marie Thompson. *Sonic Cyberfeminisms*. Goldsmiths Press, 2027.

Gonzalez, Maya. "The Gendered Circuit: Reading The Arcane of Reproduction." *Viewpoint Magazine*, 29 September 2013. https://viewpointmag.com/2013/09/28/the-gendered-circuit-reading-the-arcane-of-reproduction/.

Good, Hannah. "A Playlist for Rest, Curated by The Nap Ministry's Tricia Hersey." *The Washington Post*, 29 April 2022. https://www.washingtonpost.com/lifestyle/2022/04/29/rest-soundtrack-tricia-hersey-nap-ministry/.

Gray, Mary L., and Siddharth Suri. *Ghost Work: How to Stop Silicon Valley from Building a New Global Underclass*. Houghton Mifflin Harcourt, 2019.

Greenwald, Brian H., and John Vickrey Van Cleve. "A Deaf Variety of the Human Race: Historical Memory, Alexander Graham Bell, and Eugenics.'" *The Journal of the Gilded Age and Progressive Era* 14, no. 1 (2015): 28–48. https://doi.org/10.1017/S1537781414000528

Grossman, Babs. "The Feminised Service Sector: From Micro to Macro Analysis." *Work Organisation, Labour and Globalisation* 6, no. 1 (2012). https://doi.org/10.13169/workorgalaboglob.6.1.0063.

Gunja, Munira Z., Evan D. Gumas, Relebohile Masitha, and Laurie C. Zephyrin. *Insights into the U.S. Maternal Mortality Crisis: An International Comparison*. Commonwealth Fund, 2024. https://doi.org/10.26099/CTHN-ST75.

Gutiérrez-Rodríguez, Encarnación. *Migration, Domestic Work and Affect: A Decolonial Approach on Value and the Feminization of Labor*. Routledge, 2010.

Haebich, Anna. *Broken Circles: Fragmenting Indigenous Families 1800–2000*. Fremantle Arts Centre Press, 2001.

Haffke, Maren. "'Adaptive Environments.'" In *Techniques of Hearing*, 1st ed., edited by Michael Schillmeier, Robert Stock, and Beate Ochsner. Routledge, 2022. https://doi.org/10.4324/9781003150763-13.

Hagood, Mack. *Hush: Media and Sonic Self-Control*. Duke University Press, 2019.

Hall, Stuart, Chas Chritcher, Tony Jefferson, John Clarke, and Brian Roberts. *Policing the Crisis: Mugging, the State, and Law and Order*. Macmillan, 1978.

Hallam, Susan and Music Education Council. *The Power of Music: A Research Synthesis of the Impact of Actively Making Music on the Intellectual, Social and Personal Development of Children and Young People*. International Music Education Research Centre (iMerc), 2015.

Hancox, Dan. "Debunking the Weird Myths about London's 24-Hour Party People." Substack newsletter. *Dan Hancox's Honor Oak Riot*, 1 October 2024. https://danhancox.substack.com/p/debunking-the-weird-myths-about-londons.

Hanley, Margot, and Shiri Azenkot. "Understanding the Use of Voice Assistants by Older Adults." Version 1. Preprint, arXiv, 2021. https://doi.org/10.48550/ARXIV.2111.01210.

Haque, Sarah. "Meditation App Calm Sacks One-Fifth of Staff." *The Guardian*, 12 August 2022. https://www.theguardian.com/lifeandstyle/2022/aug/12/meditation-app-calm-sacks-one-fifth-off-staff.

Haraway, Donna J. "A Cyborg Manifesto: Science, Technology, and Socialist-Feminism in the Late Twentieth Century." In *Simians, Cyborgs, and Women The Reinvention of Nature*. Routledge, 1991.

*Harvard Law Review*. "'Regulating Eugenics." Family Law 121, no. 6 (2008): 1578–99.

Heckman, James. "Promoting Social Mobility." *Boston Review*, 2012. https://www.bostonreview.net/forum/promoting-social-mobility-james-heckman/.

Heckman, James J., and Dimitriy V. Masterov. "The Productivity Argument for Investing in Young Children." *Review of Agricultural Economics* 29, no. 3 (2007): 446–93. https://doi.org/10.1111/j.1467-9353.2007.00359.x.

Herring, Chris. "Complaint-Oriented Policing: Regulating Homelessness in Public Space." *American Sociological Review* 84, no. 5 (2019): 769–800. https://doi.org/10.1177/0003122419872671.

Hesmondhalgh, David. "User-Generated Content, Free Labour and the Cultural Industries." *Ephemera* 10, no. 3/4 (2010): 18.

Hesmondhalgh, David, and Leslie M. Meier. "What the Digitalisation of Music Tells Us about Capitalism, Culture and the Power of the Information Technology Sector." *Information, Communication & Society* 21, no. 11 (2018): 1555–70. https://doi.org/10.1080/1369118X.2017.1340498.

Hess, Juliet. "Interrupting the Symphony: Unpacking the Importance Placed on Classical Concert Experiences." *Music Education Research* 20, no. 1 (2018): 11–21. https://doi.org/10.1080/14613808.2016.1202224.

Hillers, Barbara. "Dialogue or Monologue? Lullabies in Scottish Gaelic Tradition." *Rannsachadh Na Gàidhlig* 2 (2002): 33–55.

Hirsch, Lily E. *Music in American Crime Prevention and Punishmen*t. University of Michigan Press, 2012.

Hissong, Samantha. "Calm's Head of Music Courtney Phillips—Future 25." 15 June 2021. https://www.rollingstone.com/pro/features/calm-courtney-phillips-future-25-1179422/.

Hissong, Samantha. *Post Malone, Ariana Grande, Shawn Mendes Will Drop Sleep Remixes*. 17 March 2021. https://www.rollingstone.com/pro/news/post-malone-shawn-mendes-ariana-grande-sleep-calm-1142845/.

Hochschild, Arlie Russel. *The Managed Heart: Commercialization of Human Feeling*. Updated ed. University of California Press, 2012.
Hong, Grace Kyungwon. *Death beyond Disavowal: The Impossible Politics of Difference*. University of Minnesota Press, 2015.
Howse, Chris. "Towards a Dealing Just and Kind: Reflections on Gaoling Homeless People and Mandatory Sentencing." *Alternative Law Journal* 25, no. 3 (2000). https://www.austlii.edu.au/cgi-bin/viewdoc/au/journals/AltLawJl/2000/42.html#fn1.
Hsu, Aurie, and Steven Kemper. "Enacting Sonic-Cyborg Performance through the Hybrid Body in Teka-Mori and Why Should Our Bodies End at the Skin?" *Leonardo Music Journal* 29 (2019): 83–87. https://doi.org/10.1162/lmj_a_01069.
Hui, Alexandra. "Lost: Thomas Edison's Mood MusicFound: New Ways of Listening." *Endeavour* 38, no. 2 (2014): 139–42. https://doi.org/10.1016/j.endeavour.2014.04.002.
Hunter, Caroline, and Judy Nixon. "Disciplining Women: Gender, Silence and Anti-Social Behaviour." *Criminal Justice Matters* 75, no. 1 (2009): 7–9. https://doi.org/10.1080/09627250802699657.
Hunter, Elizabeth K. "'To Cause Sleepe Safe and Shure': Dangerous Substances, Sleep Medicine and Poison Theories in Early Modern England." *Social History of Medicine* 35, no. 2 (2022): 473–93. https://doi.org/10.1093/shm/hkab064.
Hutchings, Rachel, Dominic Carter, and Katie Bennett. *Dementia—the True Cost: Fixing the Care Crisis*. 980. Alzheimer's Society, 2018.
Hutton, Elizabeth. "'Dangerous' Home Dopplers Should Be Banned." *Huffington Post*, 20 June 2017. https://www.huffingtonpost.co.uk/elizabeth-hutton/home-dopplers_b_17206242.html.
Hyman, Paula E. "Immigrant Women and Consumer Protest: The New York City Kosher Meat Boycott of 1902." *American Jewish History* 70, no. 1 (1980): 91–105.
Ilari, Beatriz. "Concerted Cultivation and Music Learning: Global Issues and Local Variations." *Research Studies in Music Education* 35, no. 2 (2013): 179–96. https://doi.org/10.1177/1321103X13509348.
Institut Marquès. "Music and Fertilisation." Accessed 12 March 2025. https://institutomarques.com/en/assisted-reproduction/music-and-fertilisation/.
Institut Marquès. Soraya Arnelas' Christmas Concert for Future Babies via the Babypod®. 29 December 2015. https://www.youtube.com/watch?v=SQNTqFRp09Q.
Institut Marquès. *Which Music Stimulates Your Baby?* 2018. https://institutomarques.com/wp-content/uploads/2018/06/0608_NDP-Estilos-Musicales-v5-EN.pdf.
International Wages for Housework Campaign. 'The International Wages for Housework Campaign [Flyer]'. undated. https://freedomarchives.org/Documents/Finder/DOC500_scans/500.020.Wages.for.Housework.pdf.

Jackson, Melissa. "Music to Deter Yobs By." *BBC News*, 10 January 2005. http://news.bbc.co.uk/1/hi/magazine/4154711.stm.

Jin, Huafeng, and Shuo Wang. Voice-based determination of physical and emotional characteristics of users. Patent US010096319B1, filed 13 March 2017, and issued 9 October 2018. https://patentimages.storage.googleapis.com/f6/a2/36/d99e36720ad953/US10096319.pdf.

Johnsen, Sarah, and Suzanne Fitzpatrick. "Revanchist Sanitisation or Coercive Care? The Use of Enforcement to Combat Begging, Street Drinking and Rough Sleeping in England." *Urban Studies* 47, no. 8 (2010): 1703–23. https://doi.org/10.1177/0042098009356128.

Johnson, Dave. "App Store Tip: Beware of Knockoffs." 1 September 2014. https://www.cbsnews.com/news/beware-of-knockoff-and-low-quality-apps-in-app-stores/.

Johnston, Caleb, and Geraldine Pratt. "Automating Adult Social Care in the UK: Extracting Value from a Crisis." *Geoforum* 151 (May 2024): 103997. https://doi.org/10.1016/j.geoforum.2024.103997.

Joint London, *Amazon Echo: Remember Baby*. 2019. https://www.youtube.com/watch?v=k6ulyhvPHUQ.

Jones, Simon C., and Thomas G. Schumacher. "Muzak: On Functional Music and Power." *Critical Studies in Mass Communication* 9, no. 2 (1992): 156–69. https://doi.org/10.1080/15295039209366822.

Kaplan, Michael. "The Self-Consuming Commodity: Audiences, Users, and the Riddle of Digital Labor." *Television & New Media* 21, no. 3 (2020): 240–59. https://doi.org/10.1177/1527476418819002.

Kassabian, Anahid. "Music for Sleeping." In *Sound, Music, Affect: Theorizing Sonic Experience*, edited by Marie Thompson and Ian Biddle. Bloomsbury, 2013.

Kassabian, Anahid. *Ubiquitous Listening: Affect, Attention, and Distributed Subjectivity*. University of California Press, 2013.

Katz, Cindi. "Accumulation, Excess, Childhood: Toward a Countertopography of Risk and Waste." *Documents d'Anàlisi Geogràfica* 57, no. 1 (2011): 47–60. https://doi.org/10.5565/rev/dag.806.

Katz, Cindi. "Childhood as Spectacle: Relays of Anxiety and the Reconfiguration of the Child." *Cultural Geographies* 15, no. 1 (2008): 5–17. https://doi.org/10.1177/1474474007085773.

Katz, Mark. *Capturing Sound: How Technology Has Changed Music*. University of California Press, 2004.

Katzman, David M. *Seven Days a Week: Women and Domestic Service in Industrializing America*. University of Illinois Press, 1981.

Keaney, Jaya. "The Racializing Womb: Surrogacy and Epigenetic Kinship." *Science, Technology, & Human Values* 47, no. 6 (2022): 1157–79. https://doi.org/10.1177/01622439211055228.

Kelion, Leo. "Amazon: How Bezos Built His Data Machine." *BBC News*. Accessed 3 September 2020. https://www.bbc.co.uk/news/extra/CLQYZENMBI/amazon-data.

Kelly, Amita. "Fact Check: Was Planned Parenthood Started to 'Control' the Black Population?" *NPR*, 14 August 2015. https://www.npr.org/sections/itsallpolitics/2015/08/14/432080520/fact-check-was-planned-parenthood-started-to-control-the-black-population.

Kenney, William Howland. *Recorded Music in American Life: The Phonograph and Popular Memory, 1890–1945*. Oxford University Press, 1999.

Kheshti, Roshanak. *Modernity's Ear: Listening to Race and Gender in World Music*. Postmillennial Pop. New York University Press, 2015.

Koenecke, Allison, Andrew Nam, Emily Lake, et al. "Racial Disparities in Automated Speech Recognition." Social Sciences. *Proceedings of the National Academy of Sciences* 117, no. 14 (2020): 7684–89. https://doi.org/10.1073/pnas.1915768117.

Kokoon, Philips x Kokoon Sleep Headphones: Now Live on Kickstarter! 2023. https://www.youtube.com/watch?v=-dgy6pfT7eY.

Koksal, Ilker. "The Sales of Smart Speakers Skyrocketed." 10 March 2020. https://www.forbes.com/sites/ilkerkoksal/2020/03/10/the-sales-of-smart-speakers-skyrocketed/?sh=79de908138ae.

Kraft, James P. "Musicians in Hollywood: Work and Technological Change in Entertainment Industries, 1926-1940." *Technology and Culture* 35, no. 2 (1994): 289. https://doi.org/10.2307/3106303.

Kruse, Holly. "Early Audio Technology and Domestic Space." *Stanford Humanities Review* 3, no. 2 (1993): 1–14.

LaBelle, Brandon. *Sonic Agency: Sound and Emergent Forms of Resistance*. Goldsmiths Press, 2018.

*Ladies' Home Journal*. "The Amberola: A New Edison Phonograph." February 1910.

Lagercrantz, Hugo, and Jean-Pierre Changeux. "The Emergence of Human Consciousness: From Fetal to Neonatal Life." *Pediatric Research* 65, no. 3 (2009): 255–60. https://doi.org/10.1203/PDR.0b013e3181973b0d.

Laing, Dave. "A Voice without a Face: Popular Music and the Phonograph in the 1890s." *Popular Music* 10, no. 1 (1991): 1–9.

Lareau, Annette. *Unequal Childhoods: Class, Race, and Family Life*. University of California Press, 2011. https://doi.org/10.1525/9780520949904.

Larrakia Nation. "App Gives Culturally Appropriate Support for Tackling Anti-Social Behaviour." Accessed 14 March 2025. https://larrakia.com/app-gives-culturally-appropriate-support-for-tackling-anti-social-behaviour/?fbclid=IwAR03b80RTU1tnoTYRrwoSoBBjfZkjGUocr-a8gLoUPegma5bAjXkEoTWejo.

Larrakia Nation. "Services." http://larrakia.com/services/.

Larrakia Nation. "The Larrakia People." Accessed 14 March 2025. http://larrakia.com/about/the-larrakia-people/.

Larson, Sarah. "Yelling at Amazon's Alexa." *The New Yorker*, 6 October 2016. https://www.newyorker.com/culture/sarah-larson/yelling-at-alexa-amazon-echo.

Learner, Sue. "Over Half of Staff Think Care Homes Should Use Alexa to Help Care for Residents." 2019. https://www.carehome.co.uk/news/article.cfm/id/1618043/quarter-care-home-staff-smart-speakers-alexa-residents.

Leber, Jessica. "Amazon Woos Advertisers with What It Knows about Consumers." *MIT Technology Review*, 21 January 2013. https://www.technologyreview.com/2013/01/21/180412/amazon-woos-advertisers-with-what-it-knows-about-consumers/.

Lees, Loretta, Hyun Bang Shin, and Ernesto López-Morales. *Planetary Gentrification*. John Wiley, 2016.

Lentjes, Rebecca. "Sounds of Life: Fetal Heartbeat Bills and the Politics of Animacy." *Sounding Out!*, 9 July 2018. https://soundstudiesblog.com/2018/07/09/sounds-of-life-fetal-heartbeat-bills-and-the-politics-of-animacy/.

Lewis, Sophie. "Amniotechnics." *The New Inquiry*, 25 January 2017. https://thenewinquiry.com/amniotechnics/.

Lewis, Sophie. *Full Surrogacy Now: Feminism against Family*. Verso, 2019.

Lippman, Alexandra. "'Law for Whom?': Responding to Sonic Illegality in Brazil's Funk Carioca." *Sound Studies* 5, no. 1 (2019): 22–36. https://doi.org/10.1080/20551940.2018.1564457.

Local Government Association. "Hampshire County Council: Pushing the Boundaries by Using Amazon Echo." 19 January 2018. https://www.local.gov.uk/hampshire-county-council-pushing-boundaries-using-amazon-echo.

Local Government and Social Care Ombudsman. "Devon County Council (17 010 182)", 8 May 2018. http://web.archive.org/web/20211020174801/https://www.lgo.org.uk/decisions/adult-care-services/assessment-and-care-plan/17-010-182#point1.

Logan, Brent. *Learning before Birth: Every Child Deserves Giftedness*. 1st Books Library, 2003.

Lords Pharmacy. "SneakPeek Heartbeat Monitor." Accessed 12 March 2025. https://lordspharmacy.net/product/sneakpeek-heartbeat-monitor.

Lynch, Jennifer K., Jon Glasby, and Suzanne Robinson. "If Telecare Is the Answer, What Was the Question? Storylines, Tensions and the Unintended Consequences of Technology-Supported Care." *Critical Social Policy* 39, no. 1 (2019): 44–65. https://doi.org/10.1177/0261018318762737.

Lynskey, Dorian. "'Alexa, Are You Invading My Privacy?'—the Dark Side of Our Voice Assistants." Technology. *The Guardian*, 9 October 2019. https://www.theguardian.com/technology/2019/oct/09/alexa-are-you-invading-my-privacy-the-dark-side-of-our-voice-assistants.

Mammoser, Gigen. "Meet the Doctor behind the World's First Vaginal Speaker." *Vice*, 20 January 2016. https://www.vice.com/en/article/we-interviewed-the-doctor-behind-the-worlds-first-vaginal-speaker/.

Manikonda, Lydia, Aditya Deotale, and Subbarao Kambhampati. "What's up with Privacy?: User Preferences and Privacy Concerns in Intelligent Personal Assistants." *Proceedings of the 2018 AAAI/ACM Conference on AI, Ethics, and Society*, 27 December 2018, 229–35. https://doi.org/10.1145/3278721.3278773.

Martin, Michèle. *"Hello, Central?": Gender, Technology, and Culture in the Formation of Telephone Systems*. McGill-Queen's University Press, 1991.

Marvin, Carolyn. *When Old Technologies Were New: Thinking about Electric Communication in the Late Nineteenth Century*. Oxford University Press, 1990.

Marx, Karl. *Capital: A Critique of Political Economy*, volume 1. Translated by Ben Fowkes. Penguin Classics. Penguin Books, 1982.

Marx, Karl, and Friedrich Engels. *The Communist Manifesto*. Edited by Jeffrey C Isaac. Yale University Press, 2017. https://doi.org/10.12987/9780300163209-007.

Mazzoni, Alana. "Government Is Accused of 'Criminalising Homelessness' after App Asks People to Dob in Rough Sleepers." *Daily Mail*. Accessed 3 February 2025. https://www.dailymail.co.uk/news/article-7313291/Government-accused-criminalising-homelessness-app-asks-people-dob-rough-sleepers.html.

Mcardle, Andrea, and Tanya Erzen. *Zero Tolerance: Quality of Life and the New Police Brutality in New York City*. New York University Press, 2001.

McCarthy, Lauren. "Feeling at Home: Between Human and AI." 8 January 2018. https://immerse.news/feeling-at-home-between-human-and-ai-6047561e7f04.

McCarthy, Lauren. "LAUREN Testimonials." 2018. https://vimeo.com/222252399?fl=pl&fe=vl.

McClintock, Anne. *Imperial Leather: Race, Gender, and Sexuality in the Colonial Contest*. Routledge, 1995.

McLennan, Chris. "New Plan to Tackle Anti-Social Behaviour Based on Darwin." *Katherine Times*, 21 October 2018. https://www.katherinetimes.com.au/story/5713857/plan-to-tackle-anti-social-behaviour/.

McLeod, Dayna. "Renovate My Uterus: Aging Queerly through Performance Art and Karaoke." *Journal of Aging Studies* 63 (2022): 100955. https://doi.org/10.1016/j.jaging.2021.100955.

McLeod, Dayna. "What a Feeling." *Journal of Autoethnography* 3, no. 2 (2022): 258–61. https://doi.org/10.1525/joae.2022.3.2.258.

Melbin, Murray. *Night as Frontier: Colonizing the World after Dark*. Free Press and Collier Macmillan, 1987. http://archive.org/details/nightasfrontiercoomelb.

Melendy, Mary Ries. *The Laws of Nature Revealed: The Science of Eugenics and Sex Life*. W. R. Vansant, 1914.

Mezzadra, Sandro, and Brett Neilson. "On the Multiple Frontiers of Extraction: Excavating Contemporary Capitalism." *Cultural Studies* 31, no. 2–3 (2017): 185–204. https://doi.org/10.1080/09502386.2017.1303425.

Michaels, Eric. *Bad Aboriginal Art: Tradition, Media, and Technological Horizons*. University of Minnesota Press, 1994. http://ebookcentral.proquest.com/lib/open/detail.action?docID=310779.

Mies, Maria. *Patriarchy and Accumulation on a World Scale: Women in the International Division of Labour*. Palgrave Macmillan, 1998.

Miles, Steven. "'Our Tyne': Iconic Regeneration and the Revitalisation of Identity in NewcastleGateshead." *Urban Studies* 42, no. 5/6 (2005): 913–26.

Miller, Karl Hagstrom. "Working Musicians: Exploring the Rhetorical Ties between Musical Labour and Leisure." *Leisure Studies* 27, no. 4 (2008): 427–41. https://doi.org/10.1080/02614360802334963.

Miller, Toby. "The Art of Waste: Contemporary Culture and Unsustainable Energy Use." In *Signal Traffic: Critical Studies of Media Infrastructures*, edited by Lisa Parks and Nicole Starosielski. University of Illinois Press, 2015.

Millie, Andrew. "Anti-Social Behaviour in British Cities: Anti-Social Behaviour in British Cities." *Geography Compass* 2, no. 5 (2008): 1681–96. https://doi.org/10.1111/j.1749-8198.2008.00135.x.

Mohandesi, Salar, and Emms Teitelman. "Without Reserves." In *Social Reproduction Theory: Remapping Class, Recentering Oppression*, edited by Tithi Bhattacharya. Pluto Press, 2017.

Monier, Mel. "'Rest as Resistance': Black Cyberfeminism, Collective Healing and Liberation on @TheNapMinistry." *Communication, Culture & Critique* 16, no. 3 (2023): 119–25. https://doi.org/10.1093/ccc/tcad022.

MonkLeading4578. "can i share my family plan with friends?" Reddit Post. R/YoutubeMusic, 2 September 2023. www.reddit.com/r/YoutubeMusic/comments/167xq3t/can_i_share_my_family_plan_with_friends/.

Morandeira, Julia. "Politics of Sleep—A Text for the Right to Rest." In *Party Studies*, volume 1, edited by Víctor Aguado, Ramón Del Buey, and Brandon LaBelle. Errant Bodies Press, 2021.

Moreton-Robinson, Aileen. *The White Possessive: Property, Power, and Indigenous Sovereignty*. University of Minnesota Press, 2015.

Morphée. Morphée Sleep Aid. Accessed 8 January 2026. https://www.morphee.co.uk/products/morphee.

Morphée, My Little Morphée. Accesed 8 January 2026. https://www.morphee.co.uk/products/my-little-morphee

Moynihan, Qayyah, and Hannah Schwär. "Amazon May Have given Alexa Female Voice to Make Device Sound 'Caring.'" *Business Insider*, 1 September 2018. https://www.businessinsider.com/theres-psychological-reason-why-amazon-gave-alexa-a-female-voice-2018-9?r=US&IR=T.

Mozilla Foundation. "Privacy Not Included: Calm." Accessed 16 January 2024. https://foundation.mozilla.org/en/privacynotincluded/calm/.

Mueller, Gavin. "The Full Automation Fallacy." *Futures of Work*, 2 December 2020. https://futuresofwork.co.uk/2020/12/02/the-full-automation-fallacy/.

Munn, Luke. "Pronatalism Is the Latest Silicon Valley Trend: What Is It—and Why Is It Disturbing?' *The Conversation*, 29 May 2024. https://doi.org/10.64628/AA.7tvwgkspf.

Murch, Walter. "FILM; Stretching Sound to Help the Mind See." *The New York Times*, 1 October 2000. https://www.nytimes.com/2000/10/01/arts/film-stretching-sound-to-help-the-mind-see.html.

Murphy, Michelle. "The Girl: Mergers of Feminism and Finance in Neoliberal Times." *The Scholar and Feminist Online* 11, nos. 1–2 (2013).

Murphy, Stacey. "'Compassionate' Strategies of Managing Homelessness: Post-Revanchist Geographies in San Francisco." *Antipode* 41, no. 2 (2009): 305–25. https://doi.org/10.1111/j.1467-8330.2009.00674.x.

"Music in the Home." *Musician* 18 no. 5 (1913): 312.

Nancy, Jean-Luc. *Listening*. Fordham University Press, 2009.

The Nap Ministry. "One Woman Shows: Transfiguration and Reparations LIVE!" August 2018. https://thenapministry.wordpress.com/2018/08/21/one-woman-shows-transfiguration-and-reparations-live/.

Nash, Jennifer C. *Birthing Black Mothers*. Duke University Press, 2021.

Negus, Keith. "From Creator to Data: The Post-Record Music Industry and the Digital Conglomerates." *Media, Culture & Society* 41, no. 3 (2019): 367–84. https://doi.org/10.1177/0163443718799395.

Newburn, Tim, and Trevor Jones. "Symbolizing Crime Control: Reflections on Zero Tolerance." *Theoretical Criminology* 11, no. 2 (2007): 221–43. https://doi.org/10.1177/1362480607075849.

NHS England Digital. "Individual Worker Characteristics." 2019. https://digital.nhs.uk/data-and-information/publications/statistical/personal-social-services-staff-of-social-services-departments/england-2019/individual-worker-characteristics#ethnicity-per-100-000-population.

Northern Media Centre. "Northern to Use Classical Music to 'Handel' Anti-Social Behaviour at Stations." 26 April 2023. https://media.northernrailway.co.uk/news/northern-to-use-classic-music-to-handel-fight-against-anti-social-behaviour-at-stations.

Northern Territory Government. "Housing." Accessed 14 March 2025. https://nteconomy.nt.gov.au/housing.

Northern Territory Government. "Population." 18 December 2025. https://nteconomy.nt.gov.au/population.

Northern Territory Government. "Reduce Crime: New Laws Introduced." Accessed 19 December 2025. https://nt.gov.au/news/2024/reduce-crime-new-laws-introduced.

NTASB. "FAQ." Accessed 14 March 2025. https://www.ntasb.com.au/faq/.

NTASB. "Understanding ASB." Accessed 14 March 2025. https://www.ntasb.com.au/understanding-asb/.

Nudelman, Franny. *Fighting Sleep: The War for the Mind and the US Military.* Verso, 2019.

Nzengou-Tayo, Marie-José. "Haitian Callaloo: What You Ask for Is Certainly Not What You Get!" *Callaloo* 30, no. 1 (2007): 175–78.

O'Brien, M. E. *Family Abolition.* 1st ed. Pluto Press, 2023.

Okigbo, Austin C. "Performing Blackness in the South African HIV/AIDS Choir." *Du Bois Review* 8, no. 1 (2011): 241–51. https://doi.org/10.1017/S1742058X11000221.

Olson, Cathy Appleford. "Meeting Of The Minds: Why The Music & Mental Wellness Industries Are Syncing Up Now More Than Ever." 18 August 2020. https://www.forbes.com/sites/cathyolson/2020/08/18/meeting-of-the-minds-why-the-music--mental-wellness-industries-are-syncing-up-now-more-than-ever/.

Oxfordshire County Council. "Care Has Just Got a Lot Smarter." 7 September 2018. https://news.oxfordshire.gov.uk/care-has-just-got-a-lot-smarter/.

PA Argenti. "Insights." https://www.paconsulting.com/insights/argenti-care-technology/.

Paine, Andre. "UMG Partners with Music Wellness App Vera for People Living with Dementia." *Music Week*, 9 June 2022. https://www.musicweek.com/digital/read/umg-partners-with-music-wellness-app-vera-for-people-living-with-dementia/085967.

Palmer Wolf, Dennie. *Lullaby: Being Together, Being Well.* WolfBrown, Weill Music Institute, 2020.

Parkhurst, Bryan. "Digital Information and Value: A Response to Jakob Rigi." *tripleC: Communication, Capitalism & Critique. Open Access Journal for a Global Sustainable Information Society* 17, no. 1 (2019): 72–85. https://doi.org/10.31269/triplec.v17i1.1078.

Parratt, Catriona M. "Making Leisure Work: Women's Rational Recreation in Late Victorian and Edwardian England." *Journal of Sport History* 26, no. 3 (1999): 471–87.

Parsell, Cameron, and Rhonda Phillips. "Indigenous Rough Sleeping in Darwin, Australia: 'Out of Place' in an Urban Setting." *Urban Studies* 51, no. 1 (2014): 185–202. https://doi.org/10.1177/0042098013484541.

Pawson, Hal, Cameron Parsell, Peter Saunders, Trish Hill, and Edgar Liu. *Australian Homelessness Monitor 2018.* 2018.

Pester, H. "Songs of Rest: An Intervention in the Complex Genre of the Lullaby." In *The Restless Compendium: Interdisciplinary Investigations of Rest and Its Opposites*, edited by Felicity Callard, Kimberley Staines, and James Wilkes. Springer International, 2016. https://doi.org/10.1007/978-3-319-45264-7_14.

Pettman, Dominic. *Sonic Intimacy: Voice, Species, Technics (or, How To Listen to the World)*. Stanford University Press, 2017.

Phan, Thao. "Amazon Echo and the Aesthetics of Whiteness." *Catalyst: Feminism, Theory, Technoscience* 5, no. 1 (2019): 1–38. https://doi.org/10.28968/cftt.v5i1.29586.

Politics Home. "Technology Can Improve Public Services but It Must Not Be a 'Codeword for Austerity,' Warns Shadow Cabinet Office Minister." 26 September 2019. https://www.politicshome.com/members/article/technology-can-improve-public-services-but-it-must-not-be-a-codeword-for-austerity-warns-shadow-cabinet-office-minister.

Pollard, Kellie. "Archaeology in the Long Grass: A Study of Aboriginal Fringe Camps in Darwin, Australia." Flinders University, 2019.

Pollard, Kellie, Claire Smith, and Jordan Ralph. "Contested Spaces: The 'Long-Grassers,' Living Private Lives in Public Places." *The Conversation*, 20 March 2017. https://doi.org/10.64628/AA.dkk4hq96v.

Povinelli, Elizabeth A. *The Cunning of Recognition: Indigenous Alterities and the Making of Australian Multiculturalism*, edited by George Steinmetz, Julia Adams, Nancy Rose Hunt, Webb Keane, and Fatma Müge Göcek. Duke University Press, 2002. https://doi.org/10.1215/9780822383673.

The Pregnancy Zone. "9 Best Prenatal Sounds and Systems for Baby and Mom." Archived 20 November 2019, at http://web.archive.org/web/20191120043121/Https://www.thepregnancyzone.com/prenatal-care/prenatal-sound-systems/.

Prey, Robert. "Musica Analytica: The Datafication of Listening." In *Networked Music Cultures*, edited by Raphaël Nowak and Andrew Whelan. Palgrave Macmillan, 2016. https://doi.org/10.1057/978-1-137-58290-4_3.

Purves, Ross M. "Local Authority Instrumental Music Tuition as a Form of Neo-Liberal Parental Investment: Findings from a Deviant, Idiographic Case Study." *Power and Education* 11, no. 3 (2019): 268–90. https://doi.org/10.1177/1757743819845068.

Radano, Ronald M. "Interpreting Muzak: Speculations on Musical Experience in Everyday Life." *American Music* 7, no. 4 (1989): 448. https://doi.org/10.2307/3051915.

Radovac, Lilian. "The 'War on Noise': Sound and Space in La Guardia's New York." *American Quarterly* 63, no. 3 (2011): 733–60. https://doi.org/10.1353/aq.2011.0038.

Rao, Smriti. "Beyond the Coronavirus: Understanding Crises of Social Reproduction." *Global Labour Journal* 12, no. 1 (2021). https://doi.org/10.15173/glj.v12i1.4307.

Raphael, Rina. "Calm's Michael Acton Smith Has Come to Hollywood to Put Audiences to Sleep." *Los Angeles Times*, 9 October 2019. https://www.latimes.com/lifestyle/story/2019-10-09/calm-comes-to-hollywood.

Rappaport, Sarah. "The Global Wellness Industry Is Now Worth $6.3 Trillion." *Bloomberg.Com*, 5 November 2024. https://www.bloomberg.com/news/articles/2024-11-05/global-wellness-industry-is-now-worth-6-3-trillion.

Rekret, Paul. *Down with Childhood: Pop Music and the Crisis of Innocence*. Repeater Books, 2017.

Rekret, Paul. "'Melodies Wander around as Ghosts': On Playlist as Cultural Form." *Critical Quarterly* 61, no. 2 (2019): 56–76. https://doi.org/10.1111/criq.12471.

Rekret, Paul. *Take This Hammer: Work, Song, Crisis*. Goldsmiths Press, 2024.

Reynolds, Michele. "Abortion Is Killing the Black Community." Ohio Senate, 18 October 2023. https://ohiosenate.gov/news/on-the-record/abortion-is-killing-the-black-community.

Reynolds, Simon. *Energy Flash: A Journey through Rave Music and Dance Culture*. Faber and Faber, 2012.

Rezaei, Mariam, and Adam Denton. "The Old Police House (TOPH): Treatise." *Contemporary Music Review* 42, no. 4 (2023): 476–87. https://doi.org/10.1080/07494467.2023.2300022.

Richardson, Sarah S. *The Maternal Imprint: The Contested Science of Maternal-Fetal Effects*. University of Chicago Press, 2021.

Richter, Max. "From Sleep: CD Boxset and Blu-Ray Audio." Accessed 6 October 2023. https://maxrichter.decca.com/products/from-sleep-cd-box-set-blu-ray-audio

Richter, Max. "Sleep." Accessed 12 March 2024. https://www.maxrichter-sleep.com/en/.

Rigi, Jakob, and Robert Prey. "Value, Rent, and the Political Economy of Social Media." *The Information Society*, 20 October 2015. World. https://www.tandfonline.com/doi/abs/10.1080/01972243.2015.1069769.

Ritchey, Marianna. *Composing Capital: Classical Music in the Neoliberal Era*. University of Chicago Press, 2019.

Ritchey, Marianna. "Musical Development: Classical Music and Gentrification in Louisville, KY." *Music & Politics* 19 (2025). https://doi.org/10.3998/mp.8033.

Riva, Michele Augusto, Vincenzo Cimino, and Stefano Sanchirico. "Gian Lorenzo Bernini's 17th Century White Noise Machine." *The Lancet Neurology* 16, no. 10 (2017): 776. https://doi.org/10.1016/S1474-4422(17)30297-1.

Roberson, Michael. "Ballroom: The Trans Sounds of Black Freedom." 2017. https://artseverywhere.ca/2017/06/16/ballroom/.

Roberts, A. O. "Echo and the Chorus of Female Machines." *Sounding Out!*, 2 March 2015. https://soundstudiesblog.com/2015/03/02/echo-and-the-chorus-of-female-machines/.

Robinson, Sandra. "Databases and Doppelgängers: New Articulations of Power'. *Configurations* 26, no. 4 (2018): 411–40. https://doi.org/10.1353/con.2018.0035.

Robson-Mainwaring, Laura. "Electro-Therapeutic Cures in the Victorian Age." Blog. The National Archives, 30 May 2022. https://webarchive

.nationalarchives.gov.uk/ukgwa/20250613143715/https://blog.nationalarchives.gov.uk/electro-therapeutic-cures-in-the-victorian-age/.

Rodgers, Tara. *Pink Noises: Women on Electronic Music and Sound.* Duke University Press, 2010.

Rollins, Judith. *Between Women: Domestics and Their Employers.* Temple University Press, 1985.

Ross, Loretta, Erika Derkas, Whitney Peoples, Lynn Roberts, and Pamela Bridgewater. *Radical Reproductive Justice: Foundation, Theory, Practice, Critique.* Feminist Press at CUNY, 2017.

Roy, Elodie. "'You Ought to See My Phonograph': The Visual Wonder of Recorded Sound (1877-1900)." In *The Making of English Popular Culture,* edited by John Storey. Routledge , 2016.

Russell, Camisha. "The Race Idea in Reproductive Technologies: Beyond Epistemic Scientism and Technological Mastery." *Journal of Bioethical Inquiry* 12, no. 4 (2015): 601–12. https://doi.org/10.1007/s11673-015-9663-3.

Salmenniemi, Suvi. "The Crisis of Social Reproduction and Therapeutic Self-Care." *The Sociological Review,* 3 January 2026, 00380261251407029. https://doi.org/10.1177/00380261251407029.

sam-page94. "spotify family plan, do they actually check you live there?" Reddit Post. R/AskUK, 29 November 2023. www.reddit.com/r/AskUK/comments/186t8m9/spotify_family_plan_do_they_actually_check_you/.

Sampson, Issy. "Spoti-Fail: Will the Streaming Service's Family Playlist Ruin Your Life?" Music. *The Guardian,* 22 October 2019. https://www.theguardian.com/music/2019/oct/22/spotify-will-the-streaming-services-family-playlist-ruin-your-life.

Sanyal, Kalyan, and Rajesh Bhattacharya. "Beyond the Factory: Globalization, Informalization of Production and the Changing Locations of Labour." In *Globalization and Labour in China and India: Impacts and Responses,* edited by Paul Bowles and John Harriss. Palgrave Macmillan UK, 2010. https://doi.org/10.1057/9780230297296_8.

Schafer, R. Murray. *The Soundscape: Our Sonic Environment and the Tuning of the World.* Simon and Schuster, 1993.

Schiller, Amy, and John McMahon. "Alexa, Alert Me When the Revolution Comes: Gender, Affect, and Labor in the Age of Home-Based Artificial Intelligence." *New Political Science* 41, no. 2 (2019): 173–91. https://doi.org/10.1080/07393148.2019.1595288.

Schuller, Kyla. *The Biopolitics of Feeling: Race, Sex, and Science in the Nineteenth Century.* Duke University Press, 2018.

Schultz, Theodore W. "Investment in Human Capital." *The American Economic Review* 51, no. 1 (1961): 1–17.

Seaver, Nick. *Computing Taste: Algorithms and the Makers of Music Recommendation.* University of Chicago Press, 2022.

Sharma, Sarah. “Going to Work in Mommy’s Basement.” *Boston Review*, 19 June 2018. http://bostonreview.net/gender-sexuality/sarah-sharma-going-work-mommys-basement.

Shelly, Edwin Taylor. “The Maternal Impression Superstition, and Why It Survives.” *Interstate Medical Journal* 12 (1905): 862–65.

Sherman, Carter. “The Rise of Pronatalism: Why Musk, Vance and the Right Want Women to Have More Babies.” *The Guardian*, 11 March 2025. https://www.theguardian.com/us-news/2025/mar/11/what-is-pronatalism-right-wing-republican.

Shuttleworth, Sally. “Fagged Out: Overwork and Sleeplessness in Victorian Professional Life.” *Interface Focus* 10, no. 3 (2020): 20190088. https://doi.org/10.1098/rsfs.2019.0088.

Skills for Care. *The State of the Adult Social Care Sector and Workforce in England. Skills for Care*, 2019.

*Sky News*. “Big Bach and Fries: McDonald’s Cuts Anti-Social Behaviour with Classical Music.” 15 March 2018. https://news.sky.com/story/big-bach-and-fries-mcdonalds-cuts-anti-social-behaviour-with-classical-music-11290515.

Sloterdijk, Peter. Bubbles, volume 1, *Spheres: Microspherology*.Semiotext(e), 2011.

Smith, Neil. “Giuliani Time: The Revanchist 1990s.” *Social Text*, no. 57 (1998): 1. https://doi.org/10.2307/466878.

Smith, Neil. “Global Social Cleansing: Postliberal Revanchism And the Export of Zero Tolerance.” *Social Justice* 28, no. 3 (85) (2001): 68–74.

Smith, Neil. “Rethinking Sleep.” *Environment and Planning D: Society and Space* 14, no. 5 (1996): 505–6. https://doi.org/10.1068/d140505.

Smith, Neil. *The New Urban Frontier: Gentrification and the Revanchist City*. Routledge, 2005.

Sofia, Zoë. “Container Technologies.” *Hypatia* 15, no. 2 (2000): 181–201.

Solanes Valldaura, Albert. “MOTHER HAS ARRIVED! Queer Kinship, Care Practices and Reconfigurations of (M)Otherhood within Ballroom Culture.” *Junctions: Graduate Journal of the Humanities* 8, no. 2 (2024): 23–36. https://doi.org/10.33391/jgjh.205.

Solie, Ruth A. *Music in Other Words: Victorian Conversations*. California Studies in 19th-Century Music 12. University of California Press, 2004. https://doi.org/10.1525/9780520930063.

Sollos. “Can Music Make You Feel Better?” Accessed 7 January 2026. https://findsollos.com/blog/musicandwellness.

Sollos. “Sollos.” Accessed 7 January 2026. https://findsollos.com.

Sollos. “Technology.” Accessed 7 January 2026. https://findsollos.com/technology.

Sonic Womb Productions. “The Sonic Womb Orrb Listening Experience at the Brain Forum 2016.” May 2016. Accessed 1 August 2023. https://sonicwomb.files.wordpress.com/2016/05/the-sonic-womb-orrb-at-the-brain-forum-leaflet.pdf.

soundBrilliance. 'soundBrilliance™'. Accessed 12 January 2024. https://www.soundbrilliance.com/.

Sousa, John Phillip. "The Menace of Mechanical Music." *Appleton's Magazine*, (1906): 278–84.

Souza, Vinicius, and Maria Eugênia Sá. 'Em São Paulo, Operação "Pancadão" Reprime Bailes Na Periferia'. *Ideias* V, no. 40 (2013): 32–33.

Sparrow, Robert. "A Not-So-New Eugenics: Harris and Savulescu on Human Enhancement." Sec. 32. *The Hastings Center Report*, 2011.

Spotify. "Introducing Spotify Kids, A New Standalone App for the Next Generation of Listeners." 30 October 2019. https://newsroom.spotify.com/2019-10-30/introducing-spotify-kids-a-new-standalone-app-for-the-next-generation-of-listeners/.

Spotify. 'Spotify Kids Is Now Available in the U.S., Canada, and France'. 31 March 2020. https://newsroom.spotify.com/2020-03-31/spotify-kids-is-now-available-in-the-u-s-canada-and-france/.

Squires, Peter, ed. *ASBO Nation: The Criminalisation of Nuisance.* Policy Press, 2012.

Srnicek, Nick. *Platform Capitalism* . . . Polity, 2019.Stabile, Carol A. 'Shooting the Mother: Fetal Photography and the Politics of Disappearance'. *Camera Obscura: Feminism, Culture, and Media Studies* 10, no. 1 (1992): 178–205. https://doi.org/10.1215/02705346-10-1_28-178.

Stakemeier, Kerstin, and Marina Vishmidt. *Reproducing Autonomy: Work, Money, Crisis and Contemporary Art.* Mute, 2016.

Stanton, Hazel M. "Report on Use of Seashore Tests at Eastman School of Music." *Music Supervisors' Journal* 12, no. 4 (1926): 20–26. https://doi.org/10.2307/3382898.

Stassen, Murray. "UMG's Latest Major AI Partnership Arrives via Tech Giant NVIDIA, with Promise of 'Antidote to Generic AI Slop.'" *Music Business Worldwide*, 6 January 2026. https://www.musicbusinessworldwide.com/umgs-latest-major-ai-partnership-arrives-via-tech-giant-nvidia-with-promise-of-antidote-to-generic-ai-slop/.

Sterne, Jonathan. *The Audible Past: Cultural Origins of Sound Reproduction.* Duke University Press, 2003.

Sterne, Jonathan. "The Theology of Sound: A Critique of Orality." *Canadian Journal of Communication*, 36, no. 1 (2011). . https://doi.org/10.22230/cjc.2011v36n2a2223.

Stone, Brad. *Amazon Unbound: Jeff Bezos and the Invention of a Global Empire*. Simon & Schuster, 2022.

Stone, Brad. "The Secret Origins of Amazon's Alexa." *Wired*, 11 May 2021. https://www.wired.com/story/how-amazon-made-alexa-smarter/.

Strengers, Yolande, and Jenny Kennedy. *The Smart Wife: Why Siri, Alexa, and Other Smart Home Devices Need a Feminist Reboot.* The MIT Press, 2020.

*Success Magazine & the National Post 1911-05: Vol 14 Iss 205*. With Internet Archive. Out-of-copyright, 1911. http://archive.org/details/sim_success-magazine-the-national-post_1911-05_14_205.

Taffel, Sy. "Escaping Attention: Digital Media Hardware, Materiality and Ecological Cost." *Culture Machine* 13 (2012).

Taussig, Michael. *Mimesis and Alterity: A Particular History of the Senses*. Routledge, 2018.

Taylor, Timothy D. "The Commodification of Music at the Dawn of the Era of 'Mechanical Music.'" *Ethnomusicology* 51, no. 2 (2007): 281–305. https://doi.org/10.2307/20174526.

Taylor, Timothy Dean, Mark Katz, and Tony Grajeda. *Music, Sound, and Technology in America: A Documentary History of Early Phonograph, Cinema, and Radio*. Duke University Press, 2012.

Terranova, Tiziana. "Free Labor." *Social Text* 18, no. 2 (2000): 33–58. https://doi.org/10.1215/01642472-18-2_63-33.

Thomas, W. Shannon, Jr. Prenatal audio apparatus. Patent US4934998, filed 10 July 1986.Thompson, Charis. *Making Parents: The Ontological Choreography of Reproductive Technologies*. MIT Press, 2007.

Thompson, Ciara. "The (Non-)Gendered Practice of Irish Traditional Lullabies." *Ethnomusicology Ireland*, no. 7 (2021): 123–36.

Thompson, E. P. "Time, Work-Discipline, and Industrial Capitalism." *Past & Present* 38, no. 1 (1967): 56–97. https://doi.org/10.1093/past/38.1.56.

Thompson, Emily Ann. *The Soundscape of Modernity: Architectural Acoustics and the Culture of Listening in America, 1900–1933*. MIT Press, 2008.

Thompson, Marie. "To Soothe or Remove? Affect, Revanchism and the Weaponized Use of Classical Music." *Communication and the Public* 2, no. 4 (2017): 272–83. https://doi.org/10.1177/2057047317741902.

Thompson, Marie. "'Your Womb, The Perfect Classroom': Prenatal Sound Systems and Uterine Audiophilia." *Feminist Review* 127, 1: (2021), pp.73–89.

Universal Music Group. "Calm and Universal Music Group Create New Sleep Mixes." 17 March 2021. https://www.universalmusic.com/calm-and-universal-music-group-create-new-sleep-remix-series/.

Universal Music Group. "Universal Music Group and Thrive Global Launch Music + Health Summit." Accessed 12 January 2024. https://www.universalmusic.com/universal-music-group-and-thrive-global-launch-music-health-summit/.

Valdez, Natali. "The Redistribution of Reproductive Responsibility: On the Epigenetics of 'Environment' in Prenatal Interventions." *Medical Anthropology Quarterly* 32, no. 3 (2018): 425–42. https://doi.org/10.1111/maq.12424.

Valle, Gaby Del. "Trad Values Meets Tech." *Political Research Associates*, 27 March 2025. https://politicalresearch.org/2025/03/27/trad-values-meets-tech.

Vicedo, Marga. *The Nature and Nurture of Love: From Imprinting to Attachment in Cold War America*. University of Chicago Press, 2014.

Vieira de Oliveira, Pedro J. S. "Weaponizing Quietness: Sound Bombs and the Racialization of Noise." *Design and Culture* 11, no. 2 (2019): 193–211. https://doi.org/10.1080/17547075.2019.1609283.

Villarreal, Jairo Matallana. "Geographies of Social Justice: The Spatial Analysis of Inequalities and Urban Change." The University of Cambridge, 2013.

Vincent, Carol. "'The Children Have Only Got One Education and You Have to Make Sure It's a Good One': Parenting and Parent–School Relations in a Neoliberal Age." *Gender and Education* 29, no. 5 (2017): 541–57. https://doi.org/10.1080/09540253.2016.1274387.

Vishmidt, Marina. "Human Capital or Toxic Asset: After the Wage." 2011. https://libcom.org/article/human-capital-or-toxic-asset-after-wage-marina-vishmidt.

Vishmidt, Marina. "The Two Reproductions in (Feminist) Art and Theory since the 1970s." *Third Text* 31, no. 1 (2017): 49–66. https://doi.org/10.1080/09528822.2017.1364331.

Voegelin, Salomé. *Sonic Possible Worlds: Hearing the Continuum of Sound*. Bloomsbury, 2014.

Vokwana, Thembela. "Sinikithemba Gospel Group and the Grassroots Struggle against HIV/AIDS Stigma in South Africa." In *Health Communication and Disease in Africa*, edited by Bankole Falade and Mercy Murire. Springer Singapore, 2021. https://doi.org/10.1007/978-981-16-2546-6_6.

Vorachek, Laura. "'The Instrument of the Century': The Piano as an Icon of Female Sexuality in the Nineteenth Century." *George Eliot: George Henry Lewes Studies*, no. 38/39 (2000): 26–43.

Walker, Gavin Robert. *Songs of the South African AIDS Crisis*. Bloomsbury Publishing USA, 2025.

Walker, Peter. "Sharp Rise in Wait Times for Perinatal Mental Health Care in England." Society. *The Guardian*, 4 September 2023. https://www.theguardian.com/society/2023/sep/04/sharp-rise-in-wait-times-for-perinatal-mental-health-care-in-england.

Walker, Rob. "Washington DC's Go-Go Music Hits Back in Fight against Gentrification." US News. *The Guardian*, 14 December 2019. https://www.theguardian.com/us-news/2019/dec/14/washington-dc-go-go-music-hits-back-gentrification-sould-of-a-city.

Walsh, Tamara, and Monica Taylor. "'You're Not Welcome Here': Police Move-On Powers and Discrimination Law." *University of New South Wales Law Journal* 30, no. 1 (2007): 151–73.

Wang, Jackie. *Carceral Capitalism*. Semiotext(e), 2018.

Warfield, Patrick. "John Philip Sousa and 'The Menace of Mechanical Music.'" *Journal of the Society for American Music* 3, no. 4 (2009): 431–63. https://doi.org/10.1017/S1752196309990678.

Watkins, Johnathan, Wahyu Wulaningsih, Charlie Da Zhou, et al. "Effects of Health and Social Care Spending Constraints on Mortality in England: A Time Trend Analysis." *BMJ Open* 7, no. 11 (2017): e017722. https://doi.org/10.1136/bmjopen-2017-017722.

Weeks, Kathi. "Life within and against Work: Affective Labour, Feminist Critique, and Post-Fordist Politics." *Ephemera* 7, no. 1 (2007): 233–49.

Weeks, Kathi. *The Problem with Work: Feminism, Marxism, Antiwork Politics, and Postwork Imaginaries*. Duke University Press, 2011.

Weheliye, Alexander G. "'Feenin.'" *Social Text* 20, no. 2 (2002): 21–47. https://doi.org/10.1215/01642472-20-2_71-21.

Weheliye, Alexander G. *Phonographies: Grooves in Sonic Afro-Modernity*. Duke University Press, 2005.

Weinbaum, Alys Eve. *The Afterlife of Reproductive Slavery: Biocapitalism and Black Feminism's Philosophy of History*. Duke University Press, 2019. https://doi.org/10.1515/9781478003281.

Weinbaum, Alys Eve. *Wayward Reproductions: Genealogies of Race and Nation in Transatlantic Modern Thought*. Duke University Press, 2004.

Weingarten, Karen. "From Maternal Impressions to Eugenics: Pregnancy and Inheritance in the Nineteenth-Century U.S." *Journal of Medical Humanities* 43, no. 2 (2022): 303–17. https://doi.org/10.1007/s10912-020-09667-x.

West, Emily. "Amazon: Surveillance as a Service." *Surveillance & Society* 17, no. 1/2 (2019): 27–33. https://doi.org/10.24908/ss.v17i1/2.13008.

Whitmarsh, Sarah. "Disinformation and the Ghost of Margaret Sanger." *Social Media + Society* 10, no. 4 (2024). https://doi.org/10.1177/20563051241290848.

Whitney, Shiloh. "Affective Indigestion: Lorde, Fanon, and Gutierrez-Rodriguez on Race and Affective Labor." *The Journal of Speculative Philosophy* 30, no. 3 (2016): 278–91. https://doi.org/10.5325/jspecphil.30.3.0278.

Wiesman, Frank H. Device for inducing sleep and rest. Patent US2659073A, filed 26 February 1952, and issued 10 November 1953. https://patents.google.com/patent/US2659073A/en.

Woods, Heather Suzanne. "Asking More of Siri and Alexa: Feminine Persona in Service of Surveillance Capitalism." *Critical Studies in Media Communication* 35, no. 4 (2018): 334–49. https://doi.org/10.1080/15295036.2018.1488082.

Wright, George. "Amazon to Pay $25m over Child Privacy Violations." Technology. *BBC News*, 31 May 2023. https://www.bbc.co.uk/news/technology-65772154.

Wright, James. *Technology in Social Care: A Review of the UK Policy Landscape*. 2020. https://doi.org/10.13140/RG.2.2.25146.41922.

Zylinska, Joanna, ed. *The Cyborg Experiments: The Extensions of the Body in the Media Age*. 1. publ. Technologies: Studies in Culture & Theory. Continuum, 2002. https://doi.org/10.5040/9781472545879.

# Index

Founded in 1893,
UNIVERSITY OF CALIFORNIA PRESS
publishes bold, progressive books and journals on topics in the arts, humanities, social sciences, and natural sciences—with a focus on social justice issues—that inspire thought and action among readers worldwide.

The UC PRESS FOUNDATION
raises funds to uphold the press's vital role as an independent, nonprofit publisher, and receives philanthropic support from a wide range of individuals and institutions—and from committed readers like you. To learn more, visit ucpress.edu/supportus.

www.ingramcontent.com/pod-product-compliance
Ingram Content Group UK Ltd.
Pitfield, Milton Keynes, MK11 3LW, UK
UKHW042026041026
476100UK00002B/4

9 780520 433649